# AN ECONOMIC
# SIMULATION MODEL
# FOR REGIONAL
# DEVELOPMENT PLANNING

# AN ECONOMIC SIMULATION MODEL FOR REGIONAL DEVELOPMENT PLANNING

by

## HERBERT H. FULLERTON
Associate Professor of Economics, Utah State University
Logan, Utah

and

## JAMES R. PRESCOTT
Professor of Economics, Iowa State University
Ames, Iowa

**ann arbor science** PUBLISHERS INC.
POST OFFICE BOX 1425 • ANN ARBOR, MICHIGAN 48106

# PREFACE

For the past decade a substantial interest in regional economic planning has presented several challenges to the social sciences. First, the conception of what constitutes an adequate informational base has been vastly extended. Though comprehensive planning has been an elusive goal in practice, the number of specialized regional planning agencies has grown and the mere replication of national economic accounting principles to subnational economic units has proven inadequate to the needs of these organizations. Second, the extended spatial and temporal basis for informational requirements has taxed the structural capacity of the simpler regional forecasting models. The lack of data at the substate regional level should stimulate interest in optimal methods of spatial disaggregation and delineation. The long-term forecasting horizon for resource-related investments extends well beyond our ability to predict structural change either in the sense of new parameters for existing equational specifications or entirely new systems of equations. Little interest has been evidenced in the problems of intra-annual forecasting, a focus of particular importance to the programs of state and local governments and an area that may present even more problems to the regional modeler than the longer term models used by the resource planner. Given the political difficulties inherent in planning for the intermediate term, it could perhaps be convincingly argued that the vast preponderance of existing econometric-forecasting models are designed for an empty set of planning constituencies.

The model presented in this study seeks to extend the informational basis for regional planning in the state of Iowa subject to the constraint of spatial and temporal consistency between the reference economy (the state) and its smaller components (counties and cities). Consistency is sought between the major economic accounts estimated from state data and broadly similar accounts disaggregated to these smaller spatial units. Six component sectors utilizing 90 computer equations constitute the complete model, and the forecasting horizon currently extends through 1985. The demographic and employment sectors provide the basis for estimating labor supply and demand by using dynamic cohort-survival and interindustry (input-output) models. The capital, labor, income and water sectors provide additional social accounting detail for the state and its subareas. To illustrate the disaggregative capabilities of the model four spatial delineations are utilized. Economic and hydrologic river basins provide the basis for resource planning within the state; economic areas (labor markets) and cities are more pertinent to the provision of multicounty and intrametropolitan governmental and private services. The accounting details for these four regionalizations are constructed from similarly consistent estimates for 120 standard areas (counties and cities) within the state for each year of the forecasting period.

The completion of this study could not have been possible without the support of numerous organizations and individuals associated with Iowa State University. The Agricultural and Home Economics Experiment Station and the Water Resources Research Institute are gratefully acknowledged for basic financial support; Drs. Karl A. Fox and John P. Mahlstede provided both personal encouragement and supplemental assistance in the final stages of the work. An intellectual debt is owed to Dr. Wilbur Maki who actively encouraged the development of large-scale regional models during his years at Iowa State and the many graduate students in the regional program who directly or indirectly contributed to this tradition. Drs. Robert Baumann, Donald Hadwiger, Neil Harl and John Timmons served on the graduate committee to which this study was originally submitted, and Betty Price and David Holmes provided invaluable programming assistance.

H. H. F.
Logan, Utah

J.R.P.
Ames, Iowa

# CONTENTS

# 1

# INTRODUCTION

## PROBLEM SETTING

The socioeconomic planning functions of subnational governments and agencies have grown substantially in the past two decades. Comprehensive river basin development, community colleges, multicounty transportation planning agencies and intrametropolitan renewal districts are but a few examples of specialized jurisdictions created to deal with particular problems. Underlying the growth of these new regional jurisdictions is a general dissatisfaction with traditional political units (states, counties and urban governments) as effective instruments of policy in dealing with problems that transcend their unchanging boundaries. These planning agencies need a vast amount of information to effectively cope with the specialized questions, assessments, priorities and jurisdictional conflicts that characterize both their short- and long-run planning objectives. A constantly improved capability to appraise current and projected socioeconomic data pertinent to the planning objectives of these agencies is crucial to their effective performance [29, 34, 35].

The current interest in land-use planning exemplifies these data needs and the extremely complicated problems of asesssing future spatial and temporal distributions of land-using activities within a region. Rural land-use studies have typically emphasized the productive characteristics of land in agricultural use, whereas urban studies have focused on commercial and residentiary attributes of a high density population. Even in urban areas very little is known about the land-absorbing characteristics of the vast number of residential and commercial units found in these areas, and a well-integrated system of land accounts for both areas is nonexistent. Projecting the future commercial use and ownership patterns of land is complicated by the numerous private and public decision processes that influence the returns to this resource. Spatial externalities, for example, are very important in urbanized areas, forming (at least in part) the rationale for zoning and building codes. The spatial distribution of industrial and residential land uses in cities has changed substantially over the past 20 years and is significantly influenced by policies undertaken in the public transportation sector. Developing models that accurately project land-use patterns is therefore complicated by the vast number of variables and causal interactions that affect these patterns, and the spatial and temporal basis on which such models are based may be fully as important as the choice of variables to include. If information is to be effectively developed and used, alternative spatial units recognizing this locational and sectoral diversity as well as the external and internal linkages among these units over time must be considered.

The problem of spatial delineation is particularly important in the development of comprehensive planning models of the type discussed in this book. The policies and programs of the regional units noted above should also be coordinated if inconsistencies are to be avoided. The economically appropriate spatial units may overlap political jurisdictions on which other policies are based and statistical regions for which consistent data are reported. Within the state of Iowa, for example, the Missouri and Mississippi Rivers form the eastern and western boundaries, placing the state within two major resource regions. Hydrologically defined river basins and labor markets overlap state boundaries on all sides of Iowa, complicating the problem of developing a method of spatial disaggregation consistent with the objectives of comprehensive planning.

An additional informational problem is the quantity and quality of data necessary for large-scale regional models. Since data systems are usually more developed for the spatially extensive political units, the ideal building block (*i.e.,* the smallest spatial unit consistent with all possible policy regions of interest) may suffer from inadequate information. Models that provide data for regional policy makers must be capable of either disaggregating data from the larger units or aggregating information up from the very smallest spatial building blocks. The quantity and quality of information versus the consistency of spatial reaggregation is often the trade-off involved in making this choice. In the model described in this book the maximum amount of spatial disaggregation is sought consistent with control totals known from publications for the largest regional unit.

Regional planning models must finally be sensitive to the major structural problems of the region's economy that are most likely to influence the policy functions of these agencies. Like many other states, Iowa shares three types of socioeconomic growth problems.

    (1) *Economic Development*

        A heavy reliance on agriculture, rapid increases in farm productivity and slow rates of increase in the demand for agricultural products have resulted in an unbalanced economic structure [19, 24, 27]. Factor substitution between capital and labor results in surplus agricultural labor that has not been adequately absorbed by economic expansion in the largest urbanized regions. Though manufacturing employment has grown steadily in recent years, fluctuations in farm income still dominate the economy and produce substantial annual changes in state revenues and expenditures. This economic dualism has resulted in very modest population increases with a substantial retention of retired families and rapid out-migration of younger adults. A model for such an economy must differentiate these major changes among producing sectors and be sensitive to patterns of net natural increase and migration along the population age spectrum.

    (2) *Human Resources*

        The disparate age distributions resulting from the major changes in economic activity summarized above present several problems in the public sector. Out-migration of young adults and declining birth rates have combined to decrease enrollments in primary schools. The inability to forecast these changes resulted in a considerable increase in school construction during the early 1960's and a substantial excess capacity at the end of that decade. Increases in the percentage of population that is retired puts a heavy burden on existing medical facilities, particularly in rural areas, and raises the public regulatory expenses associated with new retirement centers and nursing homes. Low income and numerous local family ties often prevent the rural family from out-migrating to more attractive retirement environments; Iowa is second only to Florida in the percentage of state population that is retired.

The skill-mix, occupational demands represented by new employment opportunities and age/income distribution of the population are important characteristics for human resource development. Matching local skills to job opportunities is one goal of the educational sector through the first 18–20 years of the life cycle. The curricula of both high schools and two-year vocational community colleges should be sensitive to the skills needed by local employers. Changes in educational content and the stock of skills lost or gained through net-migration also influences the array of formal courses offered in publicly supported educational institutions. Characteristics of accumulated wealth and income are important in determining how families may be able to cope with the expenses of retirement. The model described below provides an array of spatially disaggregated data pertinent to human resource planning including income, age and occupational distributions, and industrial/employment profiles. Some of the unresolved modeling problems in the human resources area are discussed in the concluding chapter.

(3) *Natural Resources*

The sectoral and spatial disaggregation of projected industry outputs and employment is an additionally desired characteristic of regional models. In the rural sector feedlots are producing animal waste disposal problems that may permanently alter the biological characteristics of downstream rivers and impoundments. Recreational uses of rivers and streams are in clear conflict with this technological change in the agricultural sector. Population growth in cities using streams as a vehicle for waste disposal shortens the potential polluting distance between the source and downstream farms and towns. The composition and level of industrial outputs produced in nodal centers influences potential air pollution problems for cities and contiguous counties; it also affects decentralized patterns of residential growth that compete with agricultural land uses on the suburban fringes. Though rural states are undergoing important changes in the composition of economic activity, both the urban and rural sectors have important problems in natural resource preservation and conservation. Models capable of dealing with these problems must distinguish settlement size and density in disaggregating projected levels of an array of economic activities.

A discussion of several resource planning efforts involving Iowa illustrates the spatial and temporal problems of model construction summarized above. Since 1962 state representatives have been engaged in two multistate comprehensive framework studies of the Upper Mississippi and Missouri River Basins. Study participants include representatives of each state in the basin and seven federal departments and agencies. The Departments of Agriculture, Army, Commerce, Health, Education and Welfare, Interior and the Federal Power Commission are represented in both groups; the Department of Labor has also been a participant in the Missouri Basin study. An array of specialized planning interests is encompassed within the water resource development effort.

Useful comparative data for state and substate regional planning are encumbered because Iowa is not wholly contained in either framework study region. Projected levels of economic activity and population, which provide the basis for estimating state and area resource requirements, are given from two different sources [50, 52]. Basic projections are provided by the National Planning Association for the Upper Mississippi region and by the Office of Business Economics for the Missouri Valley portion of the state. Although the objectives and desired projective detail of both studies are essentially similar, inconsistencies in the projected control totals and projection methodology preclude useful comparisons for subareas within Iowa not treated in the same water basin. A serious weakness is also inherent in these estimates because the methods employed did not permit a consideration of the constraining influence that limited resources imposed on patterns of economic development within subareas of the state. Finally, there is no structural model to account for the feedback effects through linkages to the Iowa economy from other basin states or regions outside of both framework studies.

To coordinate studies within both basins, a state resource planning effort was undertaken in 1963. Representatives from 12 state-affiliated commissions, agencies and educational institutions came together and formed an organization known as the State Coordinating Group for Water and Related Land Resources Planning. Though their efforts have been directed mainly in support of the two basin studies, three separate research reports were completed. A study by Maki [32] was designed to provide two sets of projected economic and demographic data for all Iowa hydrologic basins. This study provided data which was comparable to the data prepared in the framework studies for the Upper Mississippi portion of Iowa, for that part of the state included in the Missouri basin. This procedure was then repeated to obtain a data series consistent with the Missouri basin study that could be applied to substate portions within the Upper Mississippi basin. This procedure eliminated the difficulties associated with making useful comparisons of economic and demographic data for Iowa hydrologic areas not contained within the same framework study areas. However, detail on water requirements and linkages external to the state economy were not provided.

Progress was made in accounting for the impact on water requirements from exogenous final demand changes in a study by Barnard [7]. This was done within a comparative statics framework by treating water requirements by industry as linear functions of estimated sectoral gross outputs; these output estimates were obtained from state level input-output models [8, 30]. Utilizing both an employment and output base, projections by sector beyond 1980 and disaggregation of water requirements to Iowa hydrologic areas were accomplished.

Although these studies went far to provide consistent data series for the state, several serious problems remained. The following were most troublesome:

(1) Regional models are usually incapable of depicting the time-path of change in water and other resource requirements for detailed subareas of river basins and states. The constraining influence of limited resource supplies within these smaller areas is infrequently accounted for.

(2) Employment data used in making substate estimates do not explicitly reflect differences in subarea competitive positions with respect to market access and proximity to factor inputs. These employment estimates for substate areas are usually not based on models accounting for the interdependence of economic sectors within and among the broader regions that comprise these areas.

(3) Even within input-output studies, the interactions between population change and economic growth usually are not specified. Capital stock, investment, incomes, occupational distributions and employment estimates for contained subareas of the overall region are similarly excluded. Spatial disaggregation is most often confined to entire river basins or Standard Metropolitan Statistical Areas, and no consistency procedures insure that estimates for the regional totals are similar to the additive sum of the subarea projections.

These planning and modeling problems are complex and require the most detailed specification of interacting sectors within a computer-based, simulation format; the planning interdependencies noted above cannot be reconciled by simplistic methods of population trend extrapolation. Despite numerous criticisms of large-scale regional models in recent years, these methods provide the only means by which consistent socioeconomic projections based on the complex realities of regional change can be constructed. The model described in the following chapters is a contribution to that tradition and represents a blend of theoretical and empirical considerations designed to combine sectoral and spatial disaggregation techniques. The result is an extremely rich variety of projective data provided for over 100 substate regions and consistent with control totals for the state. The latter is a recognized political unit within which planning *de facto* occurs and, though resource problems often transcend state boundaries, the structure of this model provides a comprehensive basis for state and local planning for a wide array of public services [35].

## STUDY OBJECTIVE

The primary objective of this study was to develop a quantitative model capable of providing projective economic and demographic information for the reference region (Iowa) and for various preselected combinations of its contained subareas. This model should provide information for any desired set of subareas consistent with the base data and projective detail for the reference region; it should also recognize the area's unique economic and demographic features together with the interrelationships and changes in these data. The model should also allow for the cumulative impact of changes occurring at the area level to be traced to the reference region; conversely, it should allow for the impact of changes initiated outside the region and at the reference level to be traced to subareas. The time-path of predetermined indicators of development for all subareas and the reference region should be estimable.

The experimental capability of the model should be designed to depict an array of indicators for the designated spatial units that could be expected to result from anticipated data or policy-related changes. As this model utilizes simulation techniques, normative optimization is not attempted. Rather than predicting what ought to be, the model describes what is, and what may transpire in the near future given the existing socioeconomic structure of the reference region. To the degree that growth can be assessed by examining projections of economic and demographic variables, indicators of development should be provided by the model. Using these guidelines, more specific study objectives were established.

1. Construct a balanced simulation model of the recursive type, which can be used to provide projective economic and demo-

graphic information for the reference region and alternative combinations of its subareas (river basins, economic areas, and critical demand areas). This objective requires the following:

> (a) development and assembly of quantitative estimates of the technical, behavioral and definitional relationships that provide the basic structure for the six component sectors of the complete model
>
> (b) establishment of a recursive sequence for logically separable components or blocks such that an iterative loop of the complete model can be repeated through time
>
> (c) development of a mechanism within the model that permits spatial disaggregation of projected reference area economic variables as a part of each loop.

2. Determine the extent to which indices of impact and development are sensitive to selected data and policy related alternatives.

3. Determine the extent to which these indices are influenced by alternative subarea delineations.

**ORGANIZATION**

These objectives largely determine the organization of the remaining sections of this study; by chapter the following topics are discussed.

Chapter 2 contains a discussion of the basic economic concepts underlying the development of the model, the spatial and temporal context for the provision of information and the general structure of the six submodels. The former is crucial to the development of regional planning models (objective 1), and the spatial and temporal context is important for estimating consistent subregional social accounts that are sensitive to alternative policy courses of action (objectives 2 and 3).

Chapter 3 provides a detailed description of the two principal submodels usually contained in regional development simulation models. The demographic sector (labor supply) utilizes cohort survival techniques to calculate the available labor force by age and sex classifications. Separate sections describe fertility, death and migration rate estimation, and the treatment of dependent, labor active and inactive age cohorts. The employment sector (labor demand) uses input-output techniques at the reference region level to account for the sectoral distribution of gross outputs. Additional sections describe the estimation of input-output coefficients, the direct purchases table and the individual components of final demand.

Chapter 4 covers the remaining four submodels, completing the discussion of objective 1 above. The capital and labor submodels estimate gross and net investments by sector and spatially disaggregate reference area employment by an adaptation of the shift-share model. Regional income estimates are provided by occupational earnings distributions, and the water sector provides calculations of total requirements implied by the level of sectoral gross output for each year of the simulation. The emphasis in this chapter is on the special techniques utilized to expand the information base (both spatially and quantitatively) beyond variables normally estimated from the demographic and employment components of the model.

Chapter 5 describes the empirical results attained for several simulation runs of the entire model. Verification procedures that compare historical data with the model's estimates are discussed in a preliminary section and the results of extending the simulation beyond the historical phase are discussed later. Characteristics of the projected data for each of the submodels are described for this base run along with some of the principal interactions among these components of the overall model. An experimental run that constrains water supplies made available to manufacturing industries between 1970 and 1980 is analyzed and compared to the base run estimates. The regional variation in various development indicators is analyzed at the end of this chapter.

Throughout the chapters problems of empirical estimation and specification are discussed together and the equations in each submodel are presented at the end of their respective sections in Chapters 3 and 4. Chapter 6 includes a summary of procedures and results in addition to a discussion of possible extensions of the model to a broader array of environmental problems. The specialized calculations underlying dependent migration and income estimation are described in an appendix.

# 2

# REGIONAL DEVELOPMENT CONCEPTS AND INFORMATION NEEDS

## INTRODUCTION

Regional simulation models typically draw on numerous economic theories of development and growth. The theories of regional trade, location, export base and sectoral development are important concepts underlying the various submodels in this study. The contributions of these concepts appear not only in the specification of structural equations but also in the estimation of exogenous variables and the disaggregation of reference region variables to smaller spatial units. Though a full theoretical integration of these concepts is not undertaken here, the flexibility of the simulation format suggests causal linkages that are empirically estimable; the first section briefly describes these theories and their influence on the structure of our model.

The spatial and temporal context for regional social accounts is dependent on the interrelationships specified in the model. The building block upon which subarea regional units are constructed is crucial to the accuracy of the spatial variables estimated for all regions in the model and the disaggregating variable must be readily estimable from data on the smallest unit. The diversity of subarea regional units utilized in this model may necessitate a similarly diverse array of theoretical concepts; the crucial development determinants of hydrologic river basins and nodal regions, such as labor markets, may be quite different. Similarly the temporal dimensions of the information system are influenced by the specification of time-dated variables in the equations, lag structures and the assumptions underlying demographic changes where aging itself influences productivity of the labor force and the demand for public services. A discussion of the spatial and temporal basis for the information system is then presented, followed by an introduction of the general structure of the model resulting from considerations outlined in the previous sections. The principal flows among submodels are discussed as a basis for the more detailed treatment in Chapters 3 and 4. Some characteristics of the model as a general tool for regional development planning are also noted.

## REGIONAL DEVELOPMENT THEORIES

Economic growth and development are terms often used to indicate some measure of improvement in the capacity to produce or the well-being of a country or region. Positive change in a single status indicator such as per capita income is frequently referred to as "growth" in mature economies and "development" in less mature regions or countries [19]. Single status indicators often depict conflicting development goals as, for example, in many rural counties where net population decline due to out-migration and/or declining birth rates accompanies per capita income increases.

Multidimensional array of development indicators provides a more comprehensive basis for resolving logical conflicts arising from a reliance on a single development indicator. In a Tinbergen policy model [21, 51] this array of indicators comprises policy targets that may include such items as the level of income and its distribution, output per worker, employment per capita, public investment per household unit, and intergovernmental transfers. Sociological indicators such as community political participation levels, educational opportunity, nutritional levels and the incidence of disease may also be included. Though confined to economic-demographic variables, the model in this study provides as output numerous status indicators consistent with the scale of information accounts estimable using simulation methods. The model's empirical output extends well beyond the status of single development indicators usually dealt with in the theories reviewed below.

Regional trade theory emphasizes the comparative advantage in the real production costs of particular commodities in a region and changes in patterns of spatial specialization due to the mobility of productive factors [9, 47]. The internal advantage may be due to a unique resource endowment and/or lower input costs due to a technical production advantage in which the relation between inputs and outputs differs significantly among regions. The technical advantages may persist in the absence of learning, and factor immobilities may result in less than optimal input mixes in producing the major commodities traded among regions. Differential advantages of the ultimately immobile factor (land) often fix the location of primary industries, and complementary inputs (*e.g.*, climate) may set limits on the region's capability to substitute factor inputs. Trade theory also emphasizes current goods and services accounts and the role of output and factor input price differentials in determining the patterns of interregional commodity movements.

Regional economic models typically incorporate nonprice determinants of external trade flows and limited coverage of factor movements among spatial units. The origins and destinations of commodity flows are not distinguished, and export growth by industrial sector is usually related to aggregate demand variables in the relevant trading area outside of the region. Technical production coefficients are estimated for the region reflecting spatial input-output advantages (or disadvantages), and imports may be related directly to regional output or income levels. The difficulty of projecting relative sectoral export prices and the internal price changes that might encourage import substitution preclude rigorous empirical tests of most regional trade theories. Factor movements may be incorporated by migrational equations estimated in the demographic sector; these movements are often responsive to endogenously estimated unemployment rates or relative income levels. Though excluded in most models, the capital sector in this study provides estimates of sectoral depreciation, gross and net investment; these variables do not, however, respond to regional differentials in capital rental rates. Similarly, land rentals are normally not estimable endogenously even in the simulation models of more urbanized areas where specific locational algorithms for new firms (by sector) are often included.

Location theory emphasizes two characteristics of regional economies that may influence the structure of simulation models. First, the costs of input procurement, production and transportation to densely populated regions determine the resource or market orientation of producing firms. As suggested above, only in the largest urban models have specific locational determinants been successfully estimated; even in these models the lack of data precludes a full accounting of the firm's total costs. Inferences regarding trade flows are made from regional employment data, but the methods of estimating export and import flows do not directly provide a rationale for new enterprise locations. Second, there is the concept of threshold demand levels and the related process of cost minimizing agglomeration of economic activities into hierarchical spatial patterns. In this study, regional units are additively aggregated into broader spatial components at the subreference region level. The larger units are, however, resource-oriented unlike the more regular city hierarchies usually discussed in the locational literature. Also no attempt has been made to estimate population and/or income levels that might be thresholds for the introduction of new services, though this would provide an interesting extension of existing regional models.

The export base and sectoral theories of regional growth provide opposing emphases on external and internal sources of economic development; the former is extensively surveyed by Andrews [2, 3], and Borts and Stein [10] provide an example of the latter. In its simplest form, localized industries are entirely dependent on the growth of firms comprising the export base of the region. Usually manufacturing, extractive and agricultural firms comprise the export base and influence the growth of household and business serving firms with local markets for their output. Sectoral commodity exports in the interindustry model provide the closest analogy to the simplest form of the export base theory; additional exogenous sources of income are purchases of goods and services by nonlocally financed governmental units. Sectoral theories emphasize differing income elasticities of demand and rates of change in labor productivity that exist in various stages of economic growth. Internal income growth, for example, results in less than proportionate employment increases in the agricultural sector, while the rate of nonagricultural output rises at more than a proportionate rate. Internally produced services substitute for imports previously needed to support activity levels in the basic industries. Simulation models offer several means of endogenously estimating effects attributable to sectoral changes in productive activity. The demographic and income sectors provide the data for population and income elasticities of demand for final consumption purchases of households. These same variables have been used in numerous studies to project the final demand expenditure components for state, local and federal governments. On the supply side less flexibility is possible since this study utilizes input-output techniques where technical coefficients are fixed and substitutability among factors is not endogenously determined. Trends in labor productivity (output per worker) are more frequently estimated exogenously.

In summary, regional simulation models normally incorporate elements of numerous theories of area development and growth. This is usually accomplished either in the specification of the structural equations or by experimenting with parameters that are important components of these theories. Simulation models have generally been least successful in providing price estimates in factor and output markets and have therefore been less capable of testing theories where these variables are important. In this study labor income estimates are available so certain sectoral relationships may be endogenously determined; extensions might include the determination of other factor prices and the estimation of sectoral production functions allowing for input substitution. A secondary weakness of existing models is the treatment of the intraregional locations of both firms and households. The use of regional shift-share coefficients in this study provides some indication of spatial comparative advantage in access to input supplies and product markets. Also the ability of the model to spatially disaggregate reference region control totals extends well beyond existing studies. A full accounting of product and population flows among standard areas within the reference region is outside the scope of this study, however, though the capability of incorporating these flows is an important goal to be achieved in future modeling efforts.

## SPATIAL INFORMATION CONTEXT

Spatial delineation is one of the most important considerations for the regional analyst since the policy implications of existing areal disparities may vary enormously depending on the basis for regional aggregation. For a large set of spatial building blocks, regional units can be constructed that, for a particular attribute, may have a variance of almost zero or an extreme variation represented by the range value of the characteristic in the data. This represents a similarly wide range of interpretations possible with any set of regional data and has resulted in considerable interest in the "ideal" spatial unit. Though discussions usually center on the behavioral aspects of delineation (in contrast to the statistical properties of a particular set of units), the choice of subreference region units used in this study was influenced by recent delineation controversies.

There is a generally increasing recognition that no single regional unit suffices for all policy purposes. Meyer [40] discusses the traditional principles of homogeneity, nodality and programming or policy unit delineation, suggesting that the latter two are really variations on the homogeneity criterion. The real problem is selecting a characteristic (or set of characteristics) for purposes of distinguishing spatial units; there are as many regional delineations as there are characteristics. The work of Fox and Kumar [14, 15] suggests, however, that a particular type of "people-oriented" spatial unit is peculiarly useful in a wide variety of regional planning contexts. The basis for defining these regions rests on the following assumptions:

(1) Essential services and a major portion of employment are provided by a central city or nodal center.

(2) The time cost of distance between the central city and residents in the outlying area defines the outer perimeter of the spatial unit.

(3) Scale economies in providing essential services determine the viability of these centers. The minimum number of persons living within the region must be large enough to capture these economies.

These principles form the basis for the Functional Economic Area (FEA) and have a variety of regional applications. The longest tolerable commuting trip of 50–60 miles by automobile defines the outer boundary of the region in most rural and metropolitan areas; the very largest metropolitan communities may be envisaged as FEAs arranged in compact clusters. These regions are homogeneous with regard to the internal organization of their residentiary activities though they may be very heterogeneous in terms of basic activity, political organization and resource base. Since residents spend most of the weekday (and nighttime) hours within this region, this spatial unit provides a logical basis for planning most public and private residentiary services. As a programming or policy region, the FEA is somewhat more limited since its boundaries do not usually coincide with currently recognized governmental or statistical reporting units. However, multicounty planning agencies are consistent with the FEA concept, and in most parts of the United States counties provide good building blocks for labor markets.

This study recognizes the diversity of planning functions within the state and the inadequacy of a single regional unit to service every public activity. Several delineations were employed including the FEA and two variations of the homogeneity criterion. The former includes the transportation planning function primarily, though the data generated for labor markets are applicable to a variety of specialized governmental functions. Two resource-oriented delineations of river basins are used to aggregate the information on a spatial basis relevant to water resource planning activities; these data are consistent with state control totals including both the Mississippi and Missouri portions of the reference region. The 99 reference region counties provide the smallest spatial units for aggregating data for labor markets and river basins. Additionally, 21 urbanized areas within the county units are distinguished so a total of 120 "standard" areas are used. A "net" county excludes that portion of the county total accounted for by an urbanized area (cities having 10,000 plus persons in 1960); the latter are termed critical demand areas. Figures 1 through 4 illustrate the spatial delineations utilized in this study.

The six hydrologically defined river basins (RBH) within the reference region are shown in Figure 1. Physical homogeneity served

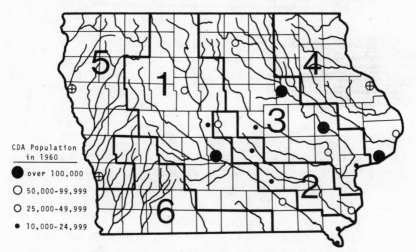

**Figure 1.** Substate hydrologic basins (RBH).

as the primary basis for their delineation as they include the major river systems within the state of Iowa. This set of regions is particularly relevant to problems dealing with surface water supplies, though in conjunction with economically defined river basins they are also appropriate spatially for considering gross waste loads and other water problems caused by highly-dense population concentrations. The RBH regions also provide a set of spatial units consistent with delineations used in the Upper Mississippi and Missouri basin framework studies [43, 52].

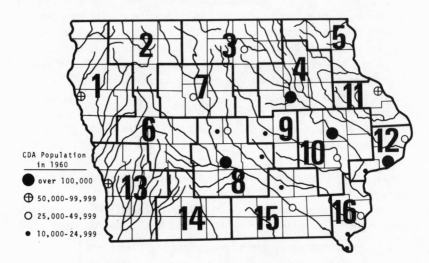

**Figure 2.**   Iowa economic areas (EA).

A second combination of standard areas as shown in Figure 2 provides an approximation to the 16 Iowa labor markets (or economic areas) as defined by Fox and Kumar [15]. Functional independence with regard to the provision of daily goods and services requirements and nodality are the major homogeneous characteristics of these regions. Since daily access to the nodal city is the basic delineating principle, this set of regions provides an especially useful context for the assessment of regional development indicators such as migration, income, employment and the centralization and/or decentralization of population within these areas.

A third set of regions (see Figure 3) was delineated by combining economic areas (EA) within substate river basins. These six regions, called economically-defined river basins (RBE), include all economic areas whose nodal city is contained within the hydrologic basin. These regions are physically homogeneous because of their relation to the hydrologic definition; additionally, they are similar in the sense that they provide the "total activity" context for assessing

the economic impact for water-related decisions affecting the hydrologic basin.

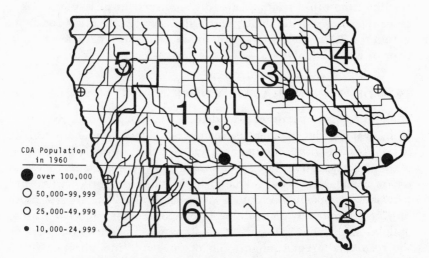

**Figure 3.** Economic river basins (RBE).

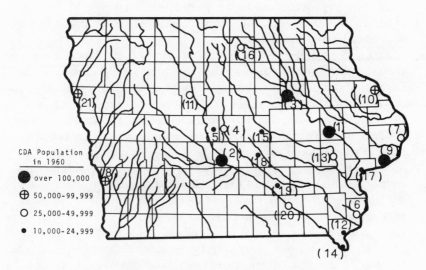

**Figure 4.** Critical demand areas (CDA).

Finally, selected standard areas (as shown in Figure 4) were sep-
arately delineated. Included among these regions are three urban
areas and 18 cities with a population exceeding 10,000 in 1960. These

regions are termed critical demand areas (CDAs) and are homogeneous in the sense of having substantially higher population densities than the other spatial units distinguished in this study. These cities typically provide the highest order services for other residents within the EA regions and are the nodal center for daily commuting trips of EA residents.

## TEMPORAL INFORMATION CONTEXT

The information context provided by considering alternative time paths of major development indicators is an important characteristic to be included within the quantitative model. Information estimated within a comparative statics framework shows the beginning and the end of change, but depicts none of the evolutionary process that lies between. At a particular date a given set of development indicators may be generated by many different temporal paths of each variable; optimal adjustment paths are as legitimate a concern to planners as the resultant target configuration of development indicators for a specified projection period. Though these considerations are not commonly included in existing regional planning models, the simulation format provides the best type of model for assessing the effects of alternative time paths of information. Several temporal features of the model are noted below.

Aging itself influences the productivity of the labor force and capital stock. In the demographic sector, the total population grows from one age cohort to the next and depreciation attributable to the utilization of the capital stock is accounted for. In each year of the simulation both the natural increase and net migration components of demographic change are estimable, unlike projections models with a fixed forecasting date or decennial "residuals" estimation techniques normally utilized to calculate net-migration by county. It should be noted that individual persons or machines are not identified, so it is not possible to estimate when specific persons entered or left the reference region or when specific capital items were purchased or retired. This would be a useful extension of present methods since it would allow modelers to identify characteristics of the gross stock (average educational attainment of the labor force, for example) at a particular point in time. In the present model aging influences a generalized population and capital stock in each year of the simulation.

The yearly simulation format also allows the modeler to date particular types of short- or long-run capacity constraints. For planning purposes it is useful, for example, to know when surface water supplies present a capacity constraint on output among river basins within the reference region. Similarly the impacts of specific time-dated resource investments on population and income can be assessed only in models allowing for information outputs at short time intervals. The sequence of planning priorities is also influenced by which types of constraints within a particular river basin are reached first. Capital stock, labor supply and numerous natural resource constraints may be reached in various sequences, and planning must be sensitive to the alternative temporal orderings of these limitations.

The recursive structure of the model depicts economic decision-making in a regular sequence of interrelationships for each time period. The decision sequence cannot be randomized or changed in a regular manner, which might reflect learning or the adaptation of reference region decision units to more efficient sequences. The stability over time of this sequence provides comparable data accounts for each year of the simulation. Also the basic structural coefficients of the principal matrix and regression linkages remain constant over time. Endogenously estimated variables may enter as independent determinants in various equations, but the structural coefficients are not reestimated on the basis of newly generated estimates of these variables. For example, moving regression linkages are not estimated and the structural input-output coefficients are not recalculated on the basis of independent estimates of the direct purchases matrix for the reference region. Though such adaptations might be a useful focus in future models seeking to study structural change within regional economies, the present information accounts are comparable over time in the sense that data are generated on the basis of an invariant set of structural relationships.

Finally, as the structural and temporal breadth of regional models expand, several adaptations can be expected. First, the estimability of new structural relationships should allow the modeler to rely less heavily on exogenous forces. Independent estimates of population and income may exert endogenous influences on the components of final demand, and relative factor returns may affect capital/labor ratios among sectors that may additionally influence import substitution. This is in accordance with the increased importance of sectoral linkages expected as a regional economy grows. Second, models with extensive temporal dimensions must be more sensitive to long term capacity constraints on growth. Natural resource depletion and the changing skills of the regional labor force are among numerous phenomena usually excluded from consideration in present projection's models, which are strongly influenced by short term theoretical considerations. For planning horizons of 20–50 years increased emphasis should be placed on structural changes internal to the model and its adaptability to long term changes in resource availability and technology.

## GENERAL STRUCTURE OF THE MODEL

The considerations discussed above influenced the overall structure of the model. The latter may be classified as a dynamic simulation model of the deterministic type [13, 37] and has the following principal characteristics:

(1) The model is decomposable into major component sectors or blocks. Each block may be simulated in isolation from other blocks as long as values for the exogenous and lagged endogenous variables can be provided; many of these variables are outputs from other component sectors of the overall model.

(2) The model is recursive in that time-lagged and sequential dependence occurs among variables within and among component sectors. Model outputs become model inputs in the recursive sequence from block to block and from one time period to the next.

(3) The model is balanced spatially. Aggregate model outputs were obtained by summation where they were generated at the standard level. Conversely, spatially disaggregated model outputs generated at the reference area level were forced to be consistent with their aggregated counterparts.

A schematic flowchart of the six-component model is shown in Figure 5. Recursive dependence is depicted by the time-dated directional arrows. For example, model output from the demographic sector in year $t$ provides an input to the labor sector in the same year; output from the latter sector is input to the demographic sector in year $t+1$. Broken lines indicate constraining relationships among sectors pertaining to capacity contraints on the capital stock and the availability of industrial water supplies. The major components of the overall model may be briefly described as follows:

(1) The demographic sector accounts for population as influenced by area, age and sex specific fertility rates, death rates and migration propensities. This sector has important linkages to the supply side of the labor sector, the final demand component of the interindustry sector, and the requirements part of the water sector.

(2) In the interindustry sector structural characteristics of the reference economy are depicted in an input-output framework.

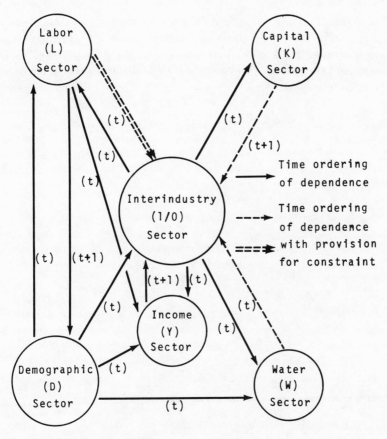

**Figure 5.** Recursive dependence among major sectors in the reference
region economy.

This model is composed of 15 interacting industries and 5 categories
of final demand with associated primary inputs, gross outlay and
gross output.

(3) The capital sector estimates gross investment, depreciation
on the capital stock, and capital accumulation; its principal linkage
is with the interindustry sector.

(4) The labor sector relates age, sex, area and occupation
specific labor supply to implied labor demand. This component is
linked primarily to the interindustry, demographic and income sectors of the overall model.

(5) The income sector provides for area, occupation and industry specific estimates of income and is linked to the interindustry,
demographic and labor sectors.

(6) The water sector determines area, industry and population specific water requirements. Level of use and disposition are also accounted for in this sector, and it uses population and output estimates provided by the demographic and interindustry sectors.

# 3

# THE DEMOGRAPHIC AND EMPLOYMENT MODELS

## INTRODUCTION

A basic characteristic of regional simulation models is the independent estimation of labor supply and demand. The supply side of the model typically accounts for the natural increase and migrational components of population change adjusted for rates of labor force participation. Labor demand is the derived demand for output estimated either from regression models differentiating the basic and nonbasic components of regional employment or more disaggregated methods such as regional input-output models. These two basic models provide estimates of the level of implied unemployment in the region. Unemployment is both an indicator of development and a variable influencing the relative attractiveness of the region to other potential inmigrating firms and households.

These two sectors are the core components of most regional development models and the subject of this chapter. The section on demographic sector describes the estimation of fertility, death and migration rates within the context of a dynamic cohort-survival model used in the demographic sector. Separate sections describe the treat-

25

ment of the dependent, labor active and inactive age cohorts and the problems of estimating cohort transition. The employment sector utilizes input-output techniques and is discussed in the next section. The estimation of the input-output table for the reference region and the components of final demand and gross output are the principal topics in this section.

## DEMOGRAPHIC SECTOR

The principal purpose of the demographic sector is to account for age and sex disaggregated population growth and spatial movements over time. Population change in a given area is assumed to be primarily influenced by the initial age-sex composition of the population, fertility, death and migration rates. Population increases (decreases) are caused by births (deaths) and positive (negative) net migration; the latter is defined as gross inmigrants minus gross outmigrants. To account for the variation in these components of population change relatively homogeneous age-sex cohorts must be determined. In this study detail is maintained on 16 age cohorts for both sexes in all 120 standard areas. The age distribution of cohorts is as follows: 0–4, 5–9, 10–14, 15–19, 20–24, 25–29, 30–34, 35–39, 40–44, 45–49, 50–54, 55–59, 60–64, 65–69, 70–74, and 75+. Additionally as the population ages, the number of persons found in the $a+1$ age group in the $t+1$ year is directly related to the number of persons found in age group $a$ in year $t$.

Estimation problems vary substantially among the principal subsectors of the demographic model. The age-sex composition of the population poses few problems since census data are available for the base year. Death rates are also readily estimated due to their relative stability over the past 25 years in both the reference region and its principal subareas. In contrast, fertility and migration rates can be quite volatile and pose numerous problems in demographic modeling. Social factors and prevailing or expected economic conditions may influence both of these components of population change. In this study fertility rates were assumed to be exogenously determined, and migration rates were affected by area unique employment prospects, industrial location and compositional variables.

A schematic flowchart of the demographic sector is shown in Figure 6. Model outputs from the labor and demographic sectors in year $t-1$ influence migration through the net employment, urbanization and employment compositional effects. Population is determined by births, deaths and migration, and the demographic estimates be-

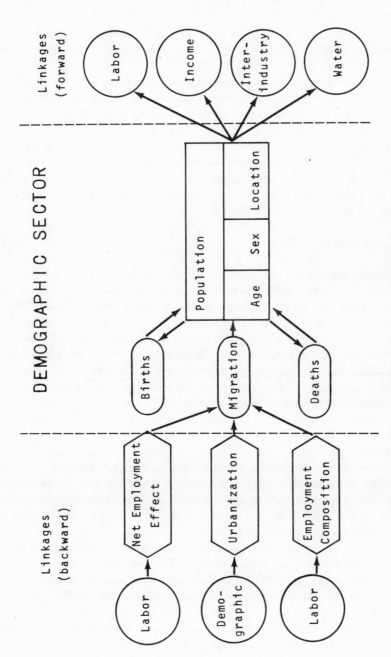

**Figure 6.** Components and sectoral linkages of the demographic sector.

come model inputs to the labor, income, interindustry and water sectors.

## Fertility Rates

Area- and age-specific fertility rates were calculated by dividing recorded live births, which were specific to the area and age of mother, by the population of females residing in the area and found in the specific age group. In the simulation runs fertility rates are not utilized until the eighth year since the actual number of live births by standard area were available for the earlier years of the period. The fertility rates calculated for the eighth year of the computer run are assumed to hold constant for the remaining years of the projection's period.

## Death Rates

Death rates were calculated in a manner similar to the estimation of fertility rates. Deaths by age cohort in each area were divided by the standard area's population found in the particular age cohort. As with live births, deaths by age cohort were read into the model as data through the year $t+8$, at which point death rates were calculated and used as constants for the remainder of the simulation period. As there are no endogenous linkages within the model, alternative assumptions regarding trends in fertility and death rates are readily incorporated into the model.

## Migration Rates

Migration rates are much more difficult to estimate due to the use of endogenous variables influencing migration and the necessity of maintaining migration detail by sex and age specific cohorts for 120 standard areas in the reference region. Two simplifications are made to facilitate the calculations.

(1) Only net migration and net migration rates are utilized primarily because of the lack of data. The use of net instead of gross migrational flows facilitated the estimation of consistent data for all standard areas by using the residual method of determining migrant flows.

(2) The population is stratified into three groups designed to reflect their respective motivations for migrating. It is assumed that young people in the dependent cohorts (1–3) migrated with

their parents or principal adult. Persons in the labor active age co-horts (4–12) are assumed to move for economic reasons, and persons exceeding 60 years in age are assumed to migrate independently of other age groups at rates observed over the period 1950–60.

*Dependent Age Groups (1–3)*

Several steps were involved in estimating dependent net migra-tion. First, it was necessary to estimate the total number of children born to migrating females during all years previous to the current year. This was done by applying area-age-sex specific fertility rates to the number of migrating females in all years previous to the cur-rent year in which they could have given birth to children whose ages in the current year would place them in the first three cohorts. For example, surviving children born between years *t-1* and *t-5* would be included in the 0–4 cohort and those born between years *t-6* and *t-10* would be in the 5–9 cohort. Females in all fertile cohorts could be expected to have children in the first cohort, but the young-est fertile cohort would not be expected to have children in the second and third cohorts. Similarly, migrating females found in co-horts beyond the childbearing age would have children only in the second or third dependent age cohorts.

A second step involved the application of survival rates to the dependent population born to migrating females to determine the number remaining in the current year. As the calculation of de-pendent net migration is based on the number of "net" female mi-grants, it is assumed that female inmigrants had identical fertility rates to those who migrated out of the area. Further, it was assumed that the children of inmigrating mothers were subject to survival rates identical to those experienced by children of the outmigrating females. Migrating children are distributed equally over the ages contained in their pertinent age cohort. These two steps are com-bined in Equation 8 shown below. The appendix describes these cal-culations in more detail.

*Labor Active Groups (4–12)*

The procedure for estimating net migration in the labor active groups represents a synthesis of several approaches to migrating households within this age group. First, migration determinants should include variables measuring prevailing or expected economic conditions. Mattila and Concannon [39] explained net migration between 1950 and 1960 on the basis of such variables as industry mix employment effect (1950–60) as a percentage of total employment (1950), the percentage of total population in age groups 20–34 years in 1950, net natural rate of change in population (1950–60), change in unemployment rates between 1950–60, and similar indicators of labor market conditions during this decade. Though generally satisfactory results were obtained, this study did not provide detailed migration estimates by age cohort, and additional constraints are imposed here due to the multisectoral structure of this model and the objective of generating the principal determinants of migration endogenously.

Second, the studies of Hamilton *et al.* [18] and Mullendore [45] used endogenously generated migrational determinants for explaining net migration by age cohort though the disaggregated independent variables generally performed less well than nonage-specific variables such as population or employment levels. Regression equations using cross section data are reevaluated at each year in the simulation run by substituting current or lagged variables estimated in other sectors of the overall model. In these models the estimates are assumed to be without error and are usually supplied from other equations in the demographic and/or employment sectors. Though it may be that the true determinants of age-specific net-migration rates are not confined to characteristics of the cohort noted above, a further problem of disaggregation is that many of the age compositional independent variables are at best spuriously correlated with the dependent variable.

In this study it is hypothesized that net migration rates for cohorts 4–12 were functionally related to the net employment effect, the proportion of employment in primary economic sectors, locational influences, and the proportion of employment in service activities. This hypothesis is summarized in Equation 7 below, and linear parameter estimates for age-sex cohorts are provided in Table I. These regressions utilized net migration rates by age-sex cohort between the years 1950 and 1960 for all standard areas within the reference region. $R^2$ for the model ranged from a low of 10% to a high of 73% for males and from a low of 17% to a high of 65% for females. All independent variables in the final equations shown in Table I are statistically significant at the 10% level or better.

Several characteristics of Table I should be noted.

(1) All intercept estimates are negative as shown in column *a* of this table.

(2) The coefficients under $b_i$ relating net migration rates to

Table I

Equation 7: Regression Coefficients and Coefficients of Determination for Estimation of Net Migration Rates by Age and Sex Groups

| Age Group | $a$ | $b_1$ | $b_2$ | $b_3$ | $b_4$ | $b_5$ | $R^2$ |
|---|---|---|---|---|---|---|---|
| **Male** | | | | | | | |
| 15-19 | -0.02449300 | 0.00002564[c] | 0.00748656[a] | 0.05094424[c] | 0.00001647[c] | -0.00000163[a] | .39822 |
| 20-24 | -0.12670000 | 0.00007797[b] | 0.04370935[c] | 0.08153293[c] | 0.00001474[c] | -0.00000147[b] | .50379 |
| 25-29 | -0.09739700 | 0.00005886[b] | 0.04606509[c] | | | | .53723 |
| 30-34 | -0.02298000 | 0.00002213[b] | 0.01852088[c] | -0.17959475[c] | | | .68711 |
| 35-39 | -0.02539500 | 0.00003208[c] | 0.01227309[c] | -0.10707031[c] | 0.00000341[b] | -0.00000077[c] | .73594 |
| 40-44 | -0.01894700 | 0.00002564[c] | 0.00670385[c] | -0.02851044[c] | 0.00000284[b] | -0.00000056[c] | .42840 |
| 45-49 | -0.01005900 | 0.00001264[c] | 0.00309314[a] | | | | .25032 |
| 50-54 | -0.01221500 | 0.00001507[c] | 0.00337356[b] | | 0.00000241[c] | -0.00000040[c] | .29144 |
| 55-59 | -0.00557730 | 0.00000851[c] | | | | | .10341 |
| **Female** | | | | | | | |
| 15-19 | -0.02827600 | 0.00003487[c] | 0.01905556[c] | 0.03434144 | 0.00001725[c] | -0.00000170[a] | .51692 |
| 20-24 | -0.11451000 | 0.00007938[b] | 0.05378618[c] | | 0.00001156[c] | -0.00000016[c] | .52821 |
| 25-29 | -0.06992300 | 0.00006406[c] | 0.02560484[c] | | | | .42500 |
| 30-34 | -0.02558900 | 0.00003891[c] | 0.00886971[b] | -0.10648543[c] | 0.00000425[b] | -0.00000106[c] | .65364 |
| 35-39 | -0.02193800 | 0.00002901[c] | 0.00523273[b] | -0.06025190[c] | 0.00000373[c] | -0.00000073[c] | .55047 |
| 40-44 | -0.01587700 | 0.00002284[c] | 0.00445029[b] | -0.01653889[c] | 0.00000248[b] | -0.00000053[c] | .31717 |
| 45-49 | -0.01181500 | 0.00001782[c] | 0.00401217[b] | | 0.00000163[b] | -0.00000036[c] | .33180 |
| 50-54 | -0.00727860 | 0.00000873[c] | 0.00246779[a] | | | | .19134 |
| 55-59 | -0.00689370 | 0.00000468[a] | 0.00221177[a] | 0.00652050[a] | | | .17174 |

[a]Significant at the 10% level.    [b]Significant at the 5% level.    [c]Significant at the 1% level.

the net employment effect are consistently positive for all age-sex cohorts.

(3) The coefficients under $b_2$, which relate the proximity of a city with a population exceeding 10,000 to net migration rates, had a positive effect for all cohorts, while $b_3$ (proximity to a major educational institution) is mixed in sign. The latter has a positive effect on males in age groups 15–19 and 20–24, and on females in the age group 15–19; a negative effect is estimated for males and females in age groups 30–34, 35–39 and 40–44. These results are consistent with urbanization trends within the reference region and the selective effect of student inmigration to standard areas with major educational institutions.

(4) Finally, $b_4$ (the proportion of employment in primary activities) and $b_5$ (changes in the proportion of employment found in service activities) had positive and negative effects respectively on net migration rates.

As incorporated in the model, estimates of area-age-sex specific net migration rates are obtained by evaluating these equations. Estimates of the independent variables are provided as model outputs from other sectors of the model. Included among these variables are the net employment effect and the primary and service employment proportions; the number of educational institutions and cities with populations exceeding 10,000 are assumed to remain unchanged. The estimated net migration rates are then multiplied by area-age-sex specific population from year *t-1* to obtain estimates of current year net migration.

*Inactive Age Groups (13–16)*

Net migration rates for this group are derived from census data for the years 1950 and 1960. In the calculation of net migration after 1960, it is assumed that the annual average rate of change would hold constant throughout the simulation period. Hence, net migration is obtained by multiplying the average one-year rate obtained from the 1950–60 period by the surviving population from the previous year.

**Cohort Transition**

In a demographic model that estimates age-specific population data, a provision must be made for aging and the transition among cohorts over time. In this study it is assumed that an equal number of persons are found within each age of a particular age cohort. This means that one-fifth of the *t-1* population in any given cohort will "age" into the next older cohort in year *t*, and each cohort in year *t* will receive one-fifth of the *t-1* population of the immediately younger cohort.

There are two exceptions to this pattern of cohort transition. First, people are born into the first cohort (0 to 4 years). Second, persons exit the terminal cohort (75+ years) by death. The proportion of one-fifth will vary depending on the time interval and age stratification utilized in a particular model. Generally, the smaller the age cohort the more accurate will be specific cohort estimates as the population ages throughout the simulation period. A larger number of age cohorts is particularly important in regional economies characterized by a substantial variance in the age distribution of the population.

## Demographic Sector Equations

Equations of the demographic sector are given below. Unless otherwise noted, the definition and range of subscripts and superscripts are as follows:

$t = 0, 1, \ldots 20$; where $t$ refers to time in years and $t=0$ in 1960

$r = 1, 2, \ldots 120$; where $r$ refers to standard areas

$s = 1, 2$; where $s$ refers to sex (1 = male, 2 = female)

$a = 1, 2, \ldots 16$; where $a$ refers to cohort (0–4, 5–9, \ldots, 70–74, 75+)

Exceptions to these ranges are noted immediately following the equation where the exception occurs.

$$(TPOP)^t = \sum_{r=1}^{120} \sum_{s=1}^{2} \sum_{a=1}^{16} (POP)^t_{rsa} \tag{1}$$

$$(POP)^t_{rsa} = (POP)^{t-1}_{rsa} - D^t_{rsa} + (NM)^t_{rsa} + (GI)^t_{rsa} - (GO)^t_{rsa}$$
$$a = 2, \ldots 16 \tag{2}$$

$$(POP)^t_{rs1} = (POP)^{t-1}_{rs1} + (LB)^t_{rs} - D^t_{rs1} + (NM)^t_{rs1} - (GO)^t_{rs1} \tag{3}$$

$$D^t_{rsa} = (POP)^{t-1}_{rs1} * (DR)^o_{rsa}$$
$$t = 8, \ldots 20 \tag{4}$$

$$(DR)^o_{rsa} = D^t_{rsa} / (POP)^t_{rsa}$$
$$t = 7 \tag{5}$$

$$(NM)^t_{rsa} = (POP)^{t-1}_{rsa} * (NMR)^t_{rsa}$$
$$a = 4, \ldots 16 \tag{6}$$

$$(NMR)^t_{rsa} = b_{0sa} + b_{1sa} * (NEE)^{t-1}_r + b_{2sa} * (DU)_r + b_{3sa} * (DE)_r$$
$$+ b_{4sa} * (EP)^{t-1}_r + b_{5sa} * (ES)^{t-1}_r$$
$$a = 4, \ldots 12 \tag{7}$$

$$(NM)^t_{rsa} = (SR)^o_{sa} * \sum_{y=a+3}^{9+a} 2.5 * (NM)^t_{r2y} * [(FR)^o_{rsy-a} + (FR)^o_{rsy-a+1}]$$
$$a = 1, 2, 3 \tag{8}$$

$$(SR)^o_{s1} = \{ 1 - [(DR)^o_{s1} * 5] \} \tag{9}$$

$$(SR)^o_{s2} = \{ 1 - [(DR)^o_{s2} * 5] * [(SR)_{s1}] \} \tag{10}$$

$$(SR)^o_{s3} = \{ 1 - [(DR)^o_{s3} * 5] * [(SR)_{s2}] \} \tag{11}$$

$$(NMR)^t_{rsa} = NMR^o_{rsa}$$
$$a = 13, \ldots 16 \tag{12}$$

$$(GL)^t_{rsa} = (POP)^{t-1}_{rsa-1}/5$$
$$a = 2, \ldots 16 \tag{13}$$

$$(GO)^t_{rsa} = (POP)^{t-1}_{tsa}/5$$
$$a = 1, \ldots 15 \tag{14}$$

$$(GO)^t_{rs16} = 0 \tag{15}$$

$$(LB)^t_{rs} = (POP)^{t-1}_{r2y} * (FR)^o_{rsy}$$
$$t = 8, \ldots 20 \tag{16}$$

$$(FR)^o_{rsy} = (LB)^t_{rsy}/(POP)^t_{r2y}$$
$$t = 7 \tag{17}$$

where:

$(TPOP)^t$ = total population, year $t$

$(POP)^t_{rsa}$ = population, year $t$, standard area $r$, sex $s$ and cohort $a$

$D^t_{rsa}$ = deaths, year $t$, standard area $r$, sex $s$ and cohort $a$

$(NM)^t_{rsa}$ = net migration, year $t$, standard area $r$, sex $s$ and cohort $a$

$(GO)^t_{rsa}$ = population leaving cohort $a$, by aging, year $t$, standard area $r$ and sex $s$

$(GI)^t_{rsa}$ = population entering cohort $a$, by aging, year $t$, standard area $r$ and sex $s$

$(LB)^t_{rsy}$ = live births, year $t$, standard area $r$, sex $s$, and to cohort of mother $y$

$(DR)^o_{rsa}$ = death rate per individual in 1967, in standard area $r$, sex $s$ and cohort $a$

(NMR)$^t_{rsa}$ = net migration rate, year $t$, standard area $r$, sex $s$ and cohort $a$

(NEE)$^t_r$ = net employment effect year $t$ in standard area $r$

(DU)$_r$ = dummy variable indicating presence of a city with population in excess of 10,000 in 1960 for standard area $r$

(DE)$_r$ = dummy variable indicating presence of a major educational institution in standard area $r$

(EP)$^t_r$ = primary employment as a proportion of total employment year $t$ and standard area $r$

(ES)$^t_r$ = service employment as a proportion of total employment year $t$ and standard area $r$

(FR)$^o_{rsy}$ = fertility rate in 1967, standard area $r$, sex of child $s$ and cohort of mother $y$

(SR)$^o_{sa}$ = survival rate in 1960, sex $s$ and cohort $a$.

## EMPLOYMENT SECTOR

There are two principal purposes for estimating employment demand by utilizing input-output techniques. (1) The interindustry model provides the best means of depicting the technical and interdependent structure of firms within the region. (2) The input-output model provides the best internally consistent means of projecting the economy of the reference region. The transactions matrix of an input-output system is composed of a set of linear production functions showing the composition of industry input requirements by industry of origin and the distribution of output by receiving industry. Projected sales to final demand (sales to households, governments, other regional industries on capital account and exports outside the region) may be converted to gross outputs demanded subject to consistent constraints imposed by purchases within the interindustry sector.

In this study the interindustry sector served to translate the impact of information output obtained from other component sectors (demographic, income, capital and water) to industries within the reference region. Similarly information output generated in the interindustry sector provided essential inputs to the labor, capital, water and income sectors. Thus, this sector accommodates numerous linkages to other component sectors in the model that are essential to the simulated recursive sequence both within and between time periods; these information linkages are depicted in Figure 7.

Numerous theoretical and empirical qualifications have been fre-

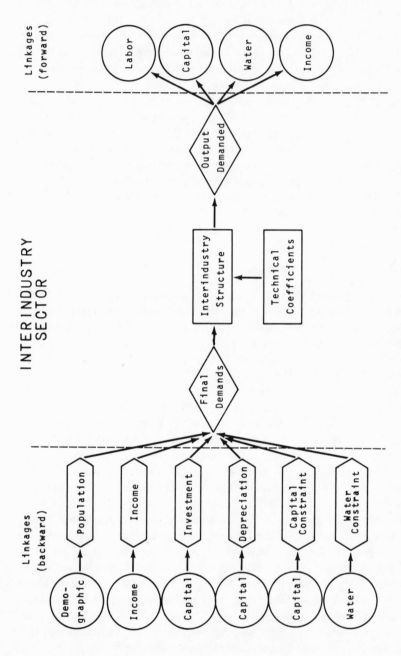

**Figure 7.** Components and sectoral linkages of the interindustry sector.

quently expressed regarding the use of regional input-output tables. Qualifications of a theoretical nature include the following: (1) assumed linearity of the production coefficients, (2) homogeneity of industrial classifications, (3) additivity of industrial products, (4) substitutability of inputs, and (5) constancy of the technical coefficients over time. The unique features of a region's industrial structure may be best captured by intensive surveys within the region though cost considerations precluded this approach for this study. The table used here is based on a large input-output table for the reference region which adapted secondary data for particularly unique features of the region's economy [30]. The extensive literature on the difficulties of applying input-output analysis to regional economies need not be elaborated on here [8, 11, 33, 44, 49].

**Input-output Tables**

Transactions, direct purchases and direct and indirect requirements matrices are estimated for the reference region economy in 1960. Both the transactions and direct and indirect requirements matrices are generated from a direct purchases matrix developed from an expanded version of the 1958 Office of Business Economics (OBE) interindustry study of the United States [17]. The methodology and data used in developing this direct purchases matrix follow Maki [30] and are discussed briefly below.

*Direct Purchases Matrix*

A direct purchases matrix shows the purchases (sales) made by a given industry from (to) all other industries for each dollar's worth of current output (outlay). The methods employed to develop this matrix relied entirely on secondary data, including the 98-sector transactions matrix for the U.S. economy and estimates of total output and total purchases of each final demand category for the reference region. First, a reference region transaction matrix was generated for producing industries by multiplying the U.S. direct requirements matrix times the vector of regional total output by industry. Estimates of each category of regional final demand were then allocated across industries in the same proportions as observed in the final demand sectors of the national transactions matrix.

At this point it was necessary to reconcile calculated regional total outputs by industry and indicated use of industry outputs to insure that transactions among regional industries and final sales were consistent with unique reference area production and consumption patterns that differ from those of the nation. This involved a comparison of the sum of intermediate purchases and final demand with total output by industry. If intermediate purchases plus final demand did not exceed regional total output for any given industry, the transactions for all elements in that industry's row were scaled downward by a factor equal to one minus the ratio of deficit to total output for that industry. Regional imports were then augmented by the extent to which their respective transactions elements were reduced. Since equality was maintained between total output and gross outlay, an adjusted direct purchases matrix including primary inputs could then be obtained for the region by dividing elements in the adjusted transactions matrix and primary input rows by the appropriate regional output and final demand figures.

These calculations were accomplished for the 102 x 98 transactions matrix for the reference region. However, to obtain industry detail consistent with other industry data and suitable for inclusion in the interindustry sector of this model, it was necessary to reduce these dimensions to an 18 x 15 matrix. This was accomplished by direct row and column summation of elements within the adjusted transactions matrix. Respective total outputs were aggregated in a similar manner to facilitate calculation of individual $a_{ij}$'s of the direct purchases matrix. The transactions matrix is shown in Table II, and the direct purchases matrix is provided in Fullerton [16, p. 62].

*Direct and Indirect Requirements Matrix*

A matrix of direct and indirect requirements shows the direct and indirect effects of demand changes that are exogenous to the processing industries. This matrix is designed to show the total expansion of output in all industries resulting from the delivery of one dollar's worth of output to final demand.

This matrix was developed for use in this study from the direct requirements matrix $A$ and an identity matrix $I$ of equal dimension. This calculation involves taking the difference $I$-$A$ between the identity matrix $I$ and the matrix $A$ and estimating the inverse matrix $(I$-$A)^{-1}$. This matrix is presented in Fullerton [16, p. 66].

**Final Demand**

Five categories are included within the final demand portion of the interindustry model. For each of these five categories a 15-element vector is estimated where each element corresponds to the industry of origin for that particular final demand category. Though all final demand categories are exogenous to the interindustry sector, only the export vector is totally exogenous to the model. Personal consumption expenditure, capital accumulation, federal government expenditure, and state and local government expenditures are calculated within the interindustry sector or are supplied to the interindustry sector as outputs from the other sectors. Base year values for final demand are obtained by using the equations shown below. Where it was not possible to evaluate the equation until the second annual iteration of the model, specific final demand categories were forced to be consistent with the estimates of Maki [30].

*Personal Consumption Expenditure*

Personal consumption expenditures (PCE) represent purchases of finished goods and services by consumers from the producing industries and are estimated by a method similar to that employed by MacMillan [28]. An estimate of total consumption per capita is calculated by inserting expected disposable income into an estimated consumption function. Next, PCE by commodity group corresponding to the 12 national account categories are estimated on the basis of regression coefficients developed by MacMillan [28]; these categories and their numerical estimates are shown in Table III. Finally, PCE by industry of origin is obtained as a product of the disbursements coefficients matrix shown in Table IV and the vector of commodity consumption expenditures. This converts consumption by commodity class into expenditure by industrial sector.

*Capital Accumulation*

Capital accumulation (CA) represents final sales by industry for domestic capital formation during the year. Annual estimates of CA are developed by calculating the product of an industry-specific investment vector and a capital coefficients matrix. This calculation translates capital expenditures by industry of purchase, as developed in the capital sector, into capital sales by industry of origin. Hence the vector of capital accumulation is exogenous to the interindustry sector, but endogenous to the model. The estimated capital coefficients matrix is shown in Table V.

## Table II
### Iowa Input-Output Table, 1960 (in $1,000)

| Industry | 1 | 2 | 3 | 4 | 5 |
|---|---|---|---|---|---|
| 1. Agriculture | 1,716,410 | 27 | 783,103 | 108 | 1,736 |
| 2. Mining and construction | 53,562 | 11,372 | 9,606 | 21 | 141 |
| 3. Food and kindred | 238,844 | 221 | 469,284 | 6 | 198 |
| 4. Textile and apparel | 706 | 9 | 2,455 | 2,844 | 156 |
| 5. Furniture, lumber and wood | 334 | 44,369 | 2,003 | 63 | 20,915 |
| 6. Printing and publishing | 986 | 141 | 5,297 | 52 | 300 |
| 7. Chemicals and allied | 35,316 | 21,371 | 12,526 | 206 | 1,493 |
| 8. Machinery | 21,044 | 18,680 | 796 | 23 | 548 |
| 9. Other and miscellaneous manufacturing | 31,892 | 200,026 | 67,624 | 1,757 | 7,375 |
| 10. Transportation | 60,288 | 26,331 | 98,886 | 426 | 3,975 |
| 11. Communication and utilities | 25,349 | 4,711 | 18,927 | 269 | 952 |
| 12. Trade | 200,425 | 92,801 | 103,619 | 1,964 | 5,573 |
| 13. Finance, insurance and real estate | 181,125 | 10,816 | 24,673 | 749 | 1,507 |
| 14. Services | 66,643 | 28,121 | 68,235 | 728 | 2,310 |
| 15. Public administration | 804 | 249 | 2,337 | 87 | 108 |
| TOTAL Intermediate Inputs | 2,633,729 | 459,245 | 1,669,372 | 9,301 | 47,285 |
| 16. Households | 1,402,257 | 263,588 | 284,671 | 13,088 | 29,444 |
| 17. Imports | 558,602 | 176,075 | 493,729 | 26,516 | 31,925 |
| 18. Other primary | 294,853 | 157,342 | 359,561 | 1,245 | 8,189 |
| TOTAL GROSS OUTLAY | 4,889,441 | 1,056,250 | 2,807,334 | 50,150 | 116,844 |

Table II, continued

| Industry | 6 | 7 | 8 | 9 | 10 |
|---|---|---|---|---|---|
| 1. Agriculture | — | 136 | 535 | 128 | 490 |
| 2. Mining and construction | 579 | 447 | 866 | 10,162 | 18,309 |
| 3. Food and kindred | — | 6,541 | 44 | 538 | 1,613 |
| 4. Textile and apparel | | 136 | 121 | 810 | 142 |
| 5. Furniture, lumber and wood | 73 | 295 | 2,855 | 4,886 | 239 |
| 6. Printing and publishing | 23,805 | 824 | 516 | 16,029 | 1,214 |
| 7. Chemicals and allied | 1,066 | 30,077 | 4,012 | 12,859 | 1,030 |
| 8. Machinery | 653 | 1,076 | 112,839 | 15,557 | 3,221 |
| 9. Other and miscellaneous manufacturing | 9,884 | 12,443 | 97,099 | 162,903 | 11,304 |
| 10. Transportation | 2,374 | 6,319 | 11,558 | 22,037 | 30,628 |
| 11. Communication and utilities | 2,486 | 3,055 | 6,828 | 13,201 | 5,407 |
| 12. Trade | 4,346 | 8,677 | 35,582 | 33,084 | 16,179 |
| 13. Finance, insurance and real estate | 7,628 | 4,435 | 10,972 | 11,707 | 25,072 |
| 14. Services | 9,298 | 14,141 | 23,853 | 17,228 | 18,442 |
| 15. Public administration | 1,309 | 687 | 1,734 | 1,215 | 10,567 |
| TOTAL Intermediate Inputs | 63,502 | 89,289 | 309,414 | 322,345 | 143,856 |
| 16. Households | 61,756 | 57,283 | 242,649 | 200,272 | 217,045 |
| 17. Imports | 43,628 | 85,783 | 289,521 | 285,624 | 111,621 |
| 18. Other primary | 13,734 | 39,589 | 82,166 | 116,069 | 77,253 |
| TOTAL GROSS OUTLAY | 182,620 | 271,944 | 923,751 | 924,309 | 549,774 |

**Table II, continued**

| Industry | 11 | 12 | 13 | 14 | 15 |
|---|---|---|---|---|---|
| 1. Agriculture | — | 2,800 | 25,220 | 827 | 8,002 |
| 2. Mining and construction | 14,520 | 14,496 | 86,025 | 17,317 | 14,926 |
| 3. Food and kindred | 8 | 10,883 | 989 | 39,110 | 7,486 |
| 4. Textile and apparel | 71 | 577 | 353 | 1,932 | 33 |
| 5. Furniture, lumber and wood | 41 | 2,143 | 342 | 226 | — |
| 6. Printing and publishing | 1,794 | 4,835 | 7,116 | 67,856 | 1,246 |
| 7. Chemicals and allied | 99 | 3,560 | 1,209 | 17,355 | 153 |
| 8. Machinery | 1,848 | 5,627 | 1,616 | 17,815 | 41 |
| 9. Other and miscellaneous manufacturing | 5,269 | 26,780 | 6,856 | 38,767 | 2,178 |
| 10. Transportation | 6,106 | 7,119 | 8,604 | 45,895 | 19,233 |
| 11. Communication and utilities | 59,824 | 50,957 | 12,944 | 42,597 | 6,991 |
| 12. Trade | 5,104 | 32,479 | 20,873 | 38,093 | 2,336 |
| 13. Finance, insurance and real estate | 6,864 | 122,941 | 153,081 | 68,547 | 2,560 |
| 14. Services | 12,473 | 108,882 | 36,799 | 79,353 | 3,645 |
| 15. Public administration | 37,663 | 26,106 | 15,666 | 8,511 | 261 |
| TOTAL Intermediate Inputs | 151,682 | 420,185 | 377,694 | 484,203 | 69,092 |
| 16. Households | 157,374 | 1,022,090 | 402,351 | 558,922 | 92,197 |
| 17. Imports | 85,963 | 282,022 | 160,786 | 217,011 | 38,998 |
| 18. Other primary | 115,467 | 231,697 | 453,090 | 13,274 | 19,530 |
| TOTAL GROSS OUTLAY | 510,486 | 1,955,994 | 1,393,920 | 1,273,410 | 180,757 |

**Table II, continued**

| Industry | Total Intermediate Demand | Total Final Demand | Total Gross Output |
|---|---|---|---|
| 1. Agriculture | 2,539,522 | 2,349,919 | 4,889,441 |
| 2. Mining and construction | 252,348 | 803,902 | 1,056,250 |
| 3. Food and kindred | 775,766 | 2,031,568 | 2,807,334 |
| 4. Textile and apparel | 10,346 | 39,804 | 50,150 |
| 5. Furniture, lumber and wood | 78,785 | 38,059 | 116,844 |
| 6. Printing and publishing | 132,010 | 50,610 | 182,620 |
| 7. Chemicals and allied | 142,331 | 129,613 | 271,944 |
| 8. Machinery | 201,384 | 722,367 | 923,751 |
| 9. Other and miscellaneous manufacturing | 682,157 | 242,152 | 924,309 |
| 10. Transportation | 349,780 | 199,994 | 549,774 |
| 11. Communication and utilities | 254,499 | 255,987 | 510,486 |
| 12. Trade | 601,134 | 1,354,860 | 1,955,994 |
| 13. Finance, insurance and real estate | 632,677 | 761,243 | 1,393,920 |
| 14. Services | 490,151 | 783,259 | 1,273,410 |
| 15. Public administration | 107,305 | 73,452 | 180,757 |
| TOTAL Intermediate Inputs | — | 9,836,790 | |
| 16. Households | 5,004,988 | 525,524 | 5,530,512 |
| 17. Imports | 2,887,804 | 1,221,488 | 4,109,302 |
| 18. Other primary | 1,943,997 | 1,103,096 | 3,047,093 |
| TOTAL GROSS OUTLAY | — | 12,686,908 | 29,773,891 |

## Table III

### Regression Coefficients for Per Capita Consumption Expenditures by Group[a]

|  | Intercept | Expenditure Elasticity | $r^2$ |
|---|---|---|---|
| 1. Food, tobacco | 4.76264 | 0.18783* | 0.90 |
| 2. Clothing | 3.24831 | 0.25966* | 0.72 |
| 3. Personal care | -5.56903 | 1.18857* | 0.95 |
| 4. Housing | -3.17559 | 1.16018* | 0.99 |
| 5. Houshold operation | 1.84577 | 0.49223* | 0.88 |
| 6. Medical care | -2.05548 | 0.88375* | 0.96 |
| 7. Personal business | -5.23677 | 1.28049* | 0.97 |
| 8. Transportation | 0.92721 | 0.59894* | 0.74 |
| 9. Recreation | -0.57170 | 0.68649* | 0.96 |
| 10. Private education | -8.92033 | 1.58609* | 0.99 |
| 11. Religion | -4.32692 | 1.00540* | 0.97 |
| 12. Foreign trade | -12.07849 | 1.94100* | 0.88 |

[a]These estimates were taken from the work of MacMillan [28].

*The coefficients are significantly different from zero at the 0.01 probability level.

### Exports

Exports comprise sales to individuals and firms located outside the reference region and are wholly exogenous to the model. This final demand category influences the level of regional gross output, but nothing within the model determines export levels. The estimates used in this study are based on the work of Maki [30]. In the simulation run annual estimates of exports by industry are obtained as a product of base year net exports and an annual growth rate as estimated by Maki [33]. These growth rates are also based on the assumption that exporting industries found in the reference region would be able to maintain a constant relative share of the external market for their products.

### Federal Government Expenditures

Federal government expenditures (FGE) are final sales to all units of federal government from industries located in the reference region. In this study FGE is assumed to be influenced by changes in

Table IV

Consumption Disbursements Coefficients, Iowa, 1960

| Industry | Food | Clothing | Personal Care | Housing |
|---|---|---|---|---|
| 1. Agriculture | 0.0537 | — | — | — |
| 2. Mining and construction | — | — | — | — |
| 3. Food and kindred | 0.5554 | — | — | — |
| 4. Textile and apparel | — | 0.3869 | 0.0003 | — |
| 5. Furniture, lumber and wood | — | — | — | — |
| 6. Printing and publishing | — | — | — | — |
| 7. Chemicals and allied | 0.0002 | — | 0.2538 | — |
| 8. Machinery | — | — | 0.0122 | — |
| 9. Other and miscellaneous manufacturing | 0.0516 | 0.1263 | 0.0344 | — |
| 10. Transportation | 0.0340 | 0.0146 | 0.0113 | — |
| 11. Communication and utilities | — | — | — | — |
| 12. Trade | 0.2947 | 0.3525 | 0.2485 | — |
| 13. Finance, insurance and real estate | — | 0.0001 | — | 0.9658 |
| 14. Services | — | 0.1127 | 0.4375 | 0.0342 |
| 15. Public administration | — | — | — | — |
| 16. Import sector | 0.0140 | 0.0069 | 0.0020 | — |
| TOTAL | 1.0000 | 1.0000 | 1.0000 | 1.0000 |

Table IV, continued

| Industry | Recreation | Private Education | Religion | Foreign Trade |
|---|---|---|---|---|
| 1. Agriculture | 0.0267 | — | — | — |
| 2. Mining and construction | — | — | — | — |
| 3. Food and kindred | — | — | — | — |
| 4. Textile and apparel | 0.0027 | — | — | — |
| 5. Furniture, lumber and wood | — | — | — | — |
| 6. Printing and publishing | 0.1347 | — | — | — |
| 7. Chemicals and allied | 0.0007 | — | — | — |
| 8. Machinery | 0.1075 | — | — | — |
| 9. Other and miscellaneous manufacturing | 0.1304 | — | — | — |
| 10. Transportation | 0.0170 | — | — | 0.1725 |
| 11. Communication and utilities | — | — | — | — |
| 12. Trade | 0.2454 | — | — | 0.1014 |
| 13. Finance, insurance and real estate | 0.0001 | — | — | — |
| 14. Services | 0.3333 | 1.0000 | 1.0000 | — |
| 15. Public administration | — | — | — | — |
| 16. Import sector | 0.0015 | — | — | 0.7261 |
| TOTAL | 1.0000 | 1.0000 | 1.0000 | 1.0000 |

Table IV, continued

| Industry | Household Operation | Medical Care | Personal Business | Transportation |
|---|---|---|---|---|
| 1. Agriculture | — | — | — | — |
| 2. Mining and construction | 0.0065 | 0.0003 | — | — |
| 3. Food and kindred | — | — | — | — |
| 4. Textile and apparel | 0.0490 | — | — | 0.0012 |
| 5. Furniture, lumber and wood | 0.0638 | — | — | — |
| 6. Printing and publishing | 0.0074 | — | — | — |
| 7. Chemicals and allied | 0.0290 | 0.0810 | — | 0.0021 |
| 8. Machinery | 0.0713 | 0.0007 | — | 0.0063 |
| 9. Other and miscellaneous manufacturing | 0.1058 | 0.0270 | — | 0.4395 |
| 10. Transportation | 0.0235 | 0.0045 | — | 0.1043 |
| 11. Communication and utilities | 0.2828 | — | 0.0015 | — |
| 12. Trade | 0.2358 | 0.1074 | — | 0.2712 |
| 13. Finance, insurance and real estate | 0.0031 | 0.0770 | 0.7820 | 0.0451 |
| 14. Services | 0.0188 | 0.7021 | 0.2083 | 0.1231 |
| 15. Public administration | 0.0151 | — | 0.0054 | 0.0069 |
| 16. Import sector | 0.0881 | — | 0.0078 | 0.0003 |
| TOTAL | 1.0000 | 1.0000 | 1.0000 | 1.0000 |

Table V

Capital Coefficients, Iowa, 1960[a,b]

| Industry | 1 | 2 | 3 | 4 | 5 |
|---|---|---|---|---|---|
| 1. Agriculture | — | — | — | — | — |
| 2. Mining and construction | 0.3500 | 0.1100 | 0.2600 | 0.1005 | 0.3000 |
| 3. Food and kindred | — | — | — | — | — |
| 4. Textile and apparel | — | 0.0001 | 0.0010 | 0.0016 | 0.0003 |
| 5. Furniture, lumber and wood | 0.0005 | 0.0003 | 0.0010 | 0.0021 | 0.0025 |
| 6. Printing and publishing | — | — | — | — | — |
| 7. Chemicals and allied | — | — | — | — | — |
| 8. Machinery | 0.4905 | 0.7124 | 0.5548 | 0.5574 | 0.4399 |
| 9. Other and miscellaneous manufacturing | 0.0222 | 0.0345 | 0.0561 | 0.1265 | 0.1525 |
| 10. Transportation | 0.1003 | 0.1186 | 0.1086 | 0.0140 | 0.0124 |
| 11. Communication and utilities | — | — | — | — | — |
| 12. Trade | 0.0341 | 0.0203 | 0.0137 | 0.0975 | 0.0883 |
| 13. Finance, insurance and real estate | — | — | — | — | — |
| 14. Services | — | — | — | — | — |
| 15. Public administration | — | — | — | — | — |

[a]Sectors 1, 3, 14 and 15 have no capital production.

[b]Source materials were obtained from Mullendore [45].

**Table V, continued**

| Industry | 6 | 7 | 8 | 9 | 10 |
|---|---|---|---|---|---|
| 1. Agriculture | — | — | — | — | — |
| 2. Mining and construction | 0.3000 | 0.2000 | 0.2721 | 0.2901 | 0.2206 |
| 3. Food and kindred | — | — | — | — | — |
| 4. Textile and apparel | 0.0010 | 0.0004 | 0.0017 | 0.0009 | 0.0020 |
| 5. Furniture, lumber and wood | 0.0010 | 0.0018 | 0.0026 | 0.0014 | 0.0150 |
| 6. Printing and publishing | — | — | — | — | — |
| 7. Chemicals and allied | — | — | — | — | — |
| 8. Machinery | 0.4720 | 0.4458 | 0.4697 | 0.4662 | 0.3050 |
| 9. Other and miscellaneous manufacturing | 0.1110 | 0.1638 | 0.1142 | 0.1074 | 0.3560 |
| 10. Transportation | 0.0210 | 0.0131 | 0.1168 | 0.0744 | 0.0720 |
| 11. Communication and utilities | — | — | — | — | — |
| 12. Trade | 0.0700 | 0.1107 | 0.0153 | 0.0388 | 0.0100 |
| 13. Finance, insurance and real estate | — | — | — | — | — |
| 14. Services | — | — | — | — | — |
| 15. Public administration | — | — | — | — | — |

Table V, continued

| Industry | 11 | 12 | 13 | 14 | 15 |
|---|---|---|---|---|---|
| 1. Agriculture | — | — | — | — | — |
| 2. Mining and construction | 0.4000 | 0.2300 | 0.3712 | 0.2254 | 0.2254 |
| 3. Food and kindred | — | — | — | — | — |
| 4. Textile and apparel | 0.0005 | 0.0040 | 0.0019 | 0.0016 | 0.0016 |
| 5. Furniture, lumber and wood | 0.0070 | 0.0670 | 0.0342 | 0.0008 | 0.0008 |
| 6. Printing and publishing | 0.0007 | 0.0098 | 0.0014 | — | — |
| 7. Chemicals and allied | 0.0024 | 0.0005 | 0.0006 | — | — |
| 8. Machinery | 0.4300 | 0.4980 | 0.1440 | 0.4759 | 0.4759 |
| 9. Other and miscellaneous manufacturing | 0.0423 | 0.0610 | 0.2514 | 0.1133 | 0.1133 |
| 10. Transportation | 0.0098 | 0.0160 | .0992 | 0.1209 | 0.1209 |
| 11. Communication and utilities | 0.0826 | — | — | — | — |
| 12. Trade | 0.0037 | 0.1090 | 0.0694 | 0.0153 | 0.0153 |
| 13. Finance, insurance and real estate | — | — | 0.0147 | — | — |
| 14. Services | — | — | — | — | — |
| 15. Public administration | — | — | — | — | — |

reference region population and other factors exogenous to the model. Annual changes in population are provided as outputs from the demographic sector and then converted to total purchases by government by an estimated growth rate of per capita FGE within the reference region. The level and distribution of FGE are based on estimates by Maki [30], and FGE growth trends are calculated on the basis of work by Mullendore [45]. Annual estimates of FGE are obtained as the product of per capita expenditure levels, total population and the annual growth rate. The distribution of sales by industry of origin is assumed to remain constant over the simulation period.

*State and Local Government Expenditures*

State and local government expenditures (SLGE) are final sales to units of state and local government from firms and individuals located in the reference region. This final demand category is assumed to be influenced by the amount of income and sales generated within the reference region and by the level of the capital stock and federal grants. The SLGE for the current year, $t$, are assumed to equal combined state and local revenues plus federal grants for the previous year, $t-1$. Observed levels of state and local revenues, including state income tax, sales tax, property taxes, and federal grants, are read into the model as data until the $t+8$ year of the simulation. After the year $t+8$ calculated sales tax, property tax and federal grant rates are used as constants throughout the remainder of the simulation run. (Federal grants are estimated as a constant proportion of revenues generated within the state.)

Tax rates estimated for use within this model may differ from those actually observed. This is due to the necessity of using base values generated within the model that may be only approximations to the actual tax base. The sales tax rate, for example, is estimated by dividing observed sales tax revenues in 1967 by gross output for that year. Also, property tax revenues in 1967 are divided by the estimated gross capital stock for the same year. In these cases gross output and capital stock serve as proxies for gross sales subject to the sales tax and taxable property valuations. As with FGE the industry proportions of SLGE are assumed to remain constant throughout the simulation period.

## Gross Output

Annual estimates of gross output are generated in the interindustry sector by multiplying the matrix of direct and indirect requirements times a vector of time-dated final demands. This calculation is shown as Equation 18.

$$(OPD)^t_i = (AINV)^o_{ij} * (FD)^t_i \tag{18}$$

where:

$(OPD)^t_i$ = vector of gross output for the $i$th industry in the year $t$

$(AINV)^o_{ij}$ = matrix of direct and indirect requirements

$(FD)^t_i$ = vector of final demands for $i$th industry in year $t$

As incorporated into the model, Equation 18 provides industry-specific upper limits on gross sales or output demanded given the existing technical structure and final demands. Since the technical structure is assumed to remain constant over the simulation period the estimation of final demands is an important step in translating consumption uses into resource employment estimates subsequently utilized in other component sectors of the model.

## Employment Sector Equations

Equations of the interindustry sector are given below. Unless otherwise noted, the definition and range of subscripts and superscripts are as follows:

$t = 0, 1, \ldots 20$; where $t$ refers to time in years and $t=0$ in 1960

$i = 1, 2, \ldots 15$; where $i$ refers to the $i$th selling industry

$j = 1, 2, \ldots 15$; where $j$ refers to the $j$th purchasing industry

$g = 1, 2, \ldots 12$; where $g$ refers to the $g$th commodity group

$o$ = refers to a base year value that remains constant

## Total Final Demand

$$(FD)^t_i = (PCE)^t_i + (CA)^t_i + (EX)^t_i + (FGE)^t_i + (SLGE)^t_i \tag{19}$$

*Personal Consumption Expenditures*

$$(PCE)^t_i = (DC)^o_{ig} * (CPC)^t_g * (TPOP)^t \tag{20}$$

$$(CPC)^t_g = EXP\,[(C)^o_g + (E)^o_g * (CPC)^t] \tag{21}$$

$$(CPC)^t = (C)^o + (PC)^o * (EYDPC)^t \tag{22}$$

*Capital Accumulation*

$$(CA)^t_i = (DK)^o_{ij} * (IVST)^t_i \tag{23}$$

*Export*

$$(EX)^t_i = [(GEX)_{ij} ** t] * (EX)^o_i \tag{24}$$

*Federal Government Expenditures*

$$(FGE)^t_i = (FGIP)^o_i * (FGE)^t \tag{25}$$

$$(FGE)^t = [(GFGE) ** t] * (FGEPC)^o * (TPOP)^t \tag{26}$$

*State and Local Government Expenditures*

$$(SLGE)^t_i = (SLIP)^o_i * (SLGR)^t \tag{27}$$

$$(SLGR)^t = K[(YTX)^t * (YP)^{t-1} + (STX)^o * (OAG)^{t-1} \\ + (PTX)^o * (KAG)^{t-1}] \\ t = 8, 9, \ldots 20 \tag{28}$$

$$(YTX)^t = R + RT * t \\ t = 8, 9, \ldots 20 \\ R = 0.00756668 \\ RT = 0.00033992 \tag{29}$$

where:

| | |
|---|---|
| $(FD)^t_i$ | = vector of total final demand for the output of industry $i$ in year $t$ |
| $(PCE)^t_i$ | = vector of personal consumption expenditures for output of industry $i$ in the year $t$ |
| $(CA)^t_i$ | = vector of capital accumulation expenditures for the output of industry $i$ in the year $t$ |
| $(EX)^t_i$ | = vector of net exports of output from industry $i$ in the year $t$ |
| $(FGE)^t_i$ | = vector of federal government expenditures for output of industry $i$ in the year $t$ |
| $(SLGE)^t_i$ | = vector of state and local governmental expenditures for output of industry $i$ in the year $t$ |
| $(DC)^o_{ig}$ | = matrix of disbursements coefficients that express the proportion of consumption for the consumption group $g$ purchased from the industry $i$ |

$(CPC)^t_g$ = vector of per capita consumption expenditures for commodity groups $g$ in the year $t$

$(TPOP)^t$ = total population, year $t$

$(C)^o_g$ = vector of constant terms in the consumption functions by commodity group $g$

$(E)^o_g$ = vector of expenditure elasticities in the consumption functions by commodity group $g$

$(CPC)^t$ = total per capita consumption expenditure in the year $t$

$(C)^o$ = constant term in the total per capita consumption function

$(PC)^o$ = constant expressing the desired proportion of personal consumption expenditure out of expected disposable income

$(EYDPC)^t$ = expected disposable income in the year $t$

$(DK)^o_{ij}$ = matrix of capital input-output coefficients expressing the proportion of industry $j$ capital purchase obtained from industry $i$

$(IVST)^t_i$ = vector of gross investment by industry $i$ in year $t$

$(GEX)_{ij}$ = diagonal matrix with nonzero elements equal to one plus the annual rate of growth in net exports from industry $i$

$(EX)^o_i$ = vector of net exports of output from industry $i$ in the base year 1960

$(FGIP)^o_i$ = vector expressing the proportion of federal government purchases from industry $i$

$(FGE)^t$ = total federal government expenditures for output produced in the reference region in year $t$

$(FGEPC)^o$ = per capita federal government expenditures for output produced in the reference region in the base year 1960

$(GFGE)$ = annual growth rate in per capita federal government expenditures

$(SLIP)^o_i$ = vector expressing the proportion of state and local government purchases from industry $i$

$(SLGR)^t$ = total state and local government revenue, including grants, in year $t$

$(YTX)^t$ = personal income tax rate for the reference area in year $t$

$(STX)^o$ = tax rate on gross sales for the reference area

$(PTX)^o$ = tax rate on estimated capital stock contained in the reference area

$(OAG)^t$ = gross sales in year $t$

$(KAG)^t$ = aggregate capital stock in year $t$

$(YP)^t$ = total personal income for the reference area in year $t$.

# 4

# THE FACTOR INPUT, INCOME AND RESOURCE MODELS

## INTRODUCTION

While the demographic and employment sectors provide the fundamental inputs to regional simulation models, numerous supplementary accounts may be estimated with the equations discussed in Chapter 3. Sectoral increases in output influence rates of net investment and may exceed existing short run constraints on capital capacity within the reference region. Temporary output increases may be partially absorbed by plants outside the reference region, and investment purchases may be made by firms within the regional economy. Labor demand is similarly derived from output increases calculated within the employment sector and is spatially disaggregated to standard areas. Estimates of the supply of labor rely on population levels calculated in the demographic sector and additional estimates of labor force participation rates and occupational profiles. These calculations complete the regional accounts for both capital and labor, which are the principal factor inputs distinguished in this model.

Measures of aggregate activity in the reference region (gross output, value added) also provide a basis for estimating labor's share of total factor payments and water resource requirements. Personal income, as an important indicator of regional development, is estimated for both the reference region economy and standard areas. Water requirements are calculated for the industrial sectors and domestic use; the former are derived from outputs of the employment sector, and domestic use rates are based on population data provided by the demographic sector. This chapter describes these four derivative submodels, completing the discussion of study objective 1.

## CAPITAL SECTOR

The objective of the capital sector is to recognize explicitly the economic linkages between sectoral outputs and investment and short run constraints on capital capacity. In any regional analysis of capital stock and investment serious data gaps inevitably limit the theoretical specification of these interrelationships and the reliability of the numerical estimates. The data used in this study are based on other studies of the reference region [6, 45] and the national economy [1, 22, 23]. This sector provides estimates of industry capital stock and investment levels, the effect of changes in capital-based technology and the restrictive effects of capital-limited output capacity. A flow chart of the capital sector (Figure 8) shows the effects of output on investment and capital with the subsequent forward linkages of depreciation and capital output capacity on the interindustry sector.

### Investment

Investment by industry provides an important data input to the final demand vector of the interindustry sector. Given annual investment estimates and a capital coefficients matrix, the static input-output model is converted to its dynamic form [20, 41] and additional projective information is provided. Investment is calculated on a gross basis at the industry level and includes depreciation on existing plant plus new investment purchases. Depreciation in the current year, $t$ is estimated as the product of a diagonal matrix of industry depreciation rates and capital stock in the year $t-1$.

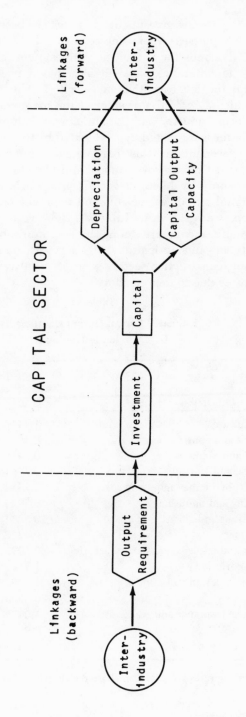

**Figure 8.** Components and sectoral linkages of the capital sector.

Current year new investment estimates are based on the following behavioral hypothesis. As businesses expand plants in anticipation of increased demands for output, new investment is assumed to equal one-half the difference between anticipated needs for capital capacity and the level existing in year $t-1$. This anticipated capital stock is calculated as the product of the incremental capital output ratio for each industry and the anticipated volume of industry output; the latter is estimated by a weighted history of lagged industrial outputs. Precedents for this treatment are found in the supply side of the Harrod-Domar growth model and two studies of the reference region economy by Maki, *et al.* [37] and MacMillan [28]. The estimated capital output coefficients and depreciation rates used in the capital sector are shown in Table VI. Relative to output levels high capital requirements are found in communications and utilities, transportation and agriculture; the services, public administration and mining/construction sectors generate relatively high investment expenditures due to depreciation.

Table VI

Estimated Capital-Output Coefficients and
Depreciation Rates, Iowa, 1960

| Industry | Capital-Output Coefficient (Ratio) | Depreciation Rate[a] (Per Cent) |
|---|---|---|
| 1. Agriculture | 1.1114 | 7.70 |
| 2. Mining and construction | 0.1909 | 9.27 |
| 3. Food and kindred | 0.2460 | 6.52 |
| 4. Textile and apparel | 0.3266 | 6.68 |
| 5. Furniture, lumber and wood | 0.2381 | 6.89 |
| 6. Printing and publishing | 0.3940 | 6.63 |
| 7. Chemicals and allied | 0.5011 | 6.66 |
| 8. Machinery | 0.4382 | 5.77 |
| 9. Other and miscellaneous manufacturing | 0.5926 | 8.19 |
| 10. Transportation | 1.7609 | 3.81 |
| 11. Communication and utilities | 3.0082 | 3.29 |
| 12. Trade | 0.6523 | 5.22 |
| 13. Finance, insurance and real estate | 1.0471 | 4.73 |
| 14. Services | 0.9451 | 9.69 |
| 15. Public administration | 0.9451 | 9.69 |

[a]Depreciation rates are based on those developed by Mullendore [45].

**Capital Stock and Capacity**

The level of an industry's capital stock in the current year $t$ is assumed to equal the capital stock in year $t-1$ plus gross investment less depreciation. Data on the capital stock by industry is used to calculate the capital-limited output capacity. This limitation is estimated by multiplying the inverse of the diagonal matrix of capital output ratios times that industry's estimated capital stock for the same year. The evaluation of this capacity limitation provides a means of simulating the impact of alternative restrictions on the growth of output due to short run capital constraints.

**Capital Sector Equations**

The equations of the capital sector are listed below. The definition and range of subscripts and superscripts are as follows:

$t=0, 1, \ldots . 20$; where $t$ refers to time in years and $t=0$ in 1960
$i=1, 2, \ldots . 15$; where $i$ refers to the $i$th selling industry
$j=1, 2, \ldots . 15$; where $j$ refers to the $j$th purchasing industry
$0=$a base year value that remains constant

$$(IVST)^t_i = (EKAP)^{t+1}_i - (KAP)^{t-1}_i + (DPR)_{ij} * (KAP)^{t-1}_i$$
(30)

$$(EKAP)^{t+1}_i = (KOP)^o_{ij} * (GOP)^t_{ij} * (OP)^{t-1}_i$$
(31)

$$(GOP)^t_{ij} = [(OP)^{t-1}_i / (OP)^{t-2}_i * 0.75] + [(OP)^{t-2}_i / (OP)^{t-3}_i * 0.25]$$
(32)

$$(KAP)^t_i = [(KAP)^{t-1}_i + (IVST)^t_i] - [(DPR)^o_{ij} * (KAP)^{t-1}_i]$$
(33)

$$(OPK)^t_i = (KOP^{-1})_{ij} * (KAP)^t_i$$
(34)

where:

$(IVST)^t_i$ = vector of gross investment by industry $i$ in the year $t$

$(DPR)^o_{ij}$ = diagonal matrix with nonzero element equal to the annual depreciation rate on capital stock in industry $i$ in year $t$

$(KAP)^t_i$ = vector of capital stock for industry $i$ in year $t$

$(EKAP)_{ti}$ = vector of expected stock in industry $i$ in year $t$

$(KOP)^o_{ij}$ = diagonal matrix with nonzero element equal to one over the rates of capital stock in industry $i$ to output for the same industry in year $t$

$(OP)^t_i$ = vector of realized gross output for industry $i$ in the year $t$

$(GOP)_{ij}$ = diagonal matrix with nonzero elements equal to one plus the annual growth rate in industry output

$(OPK)^t_i$ = vector with elements expressing the capital-limited output capacity for industry $i$ in year $t$.

## LABOR SECTOR

In this sector estimates of the available labor force (supply) and the required labor force (demand) are calculated. Changes in these estimates are due primarily to variables calculated in the demographic and interindustry sectors. Figure 9 provides a flow diagram for the principal linkages in the labor sector that are discussed below.

The estimation techniques utilized in this sector allow the time-path of labor supply by age, sex, occupation and standard area to be traced throughout the simulation period. The Census of Population major occupational groups are used including professional-technical, managers, clerical, sales, craftsmen and foremen, operatives, service workers and laborers. A similar provision is made for the labor demand variables with the exception that these estimates are identified by industry and not by age-sex group.

### Labor Supply

Labor supply estimates are based on population levels, labor force participation rates and occupational profiles. Within each year of the simulation, area-age-sex specific estimates of the available labor force are obtained, First, labor participation rates are updated to the current year by multiplying annual rates of change in labor force participation times the participation rate in the base year. Then age-sex groupings from the demographic model are combined into age-sex cohorts more pertinent to the labor sector; these groupings include age groups 14–17, 18–24, 25–34, 35–44, 45–64 and 65+. Finally, the available labor force is obtained as the product of these subgroups of the population times their respective labor participation rate. A disaggregation of the labor force for the eight occupational groups is obtained by multiplying it times the base year occupational profile for the labor force. As shown in Figure 9 the aggregate labor supply estimate determines (in conjunction with the required labor force) the level of implied unemployment.

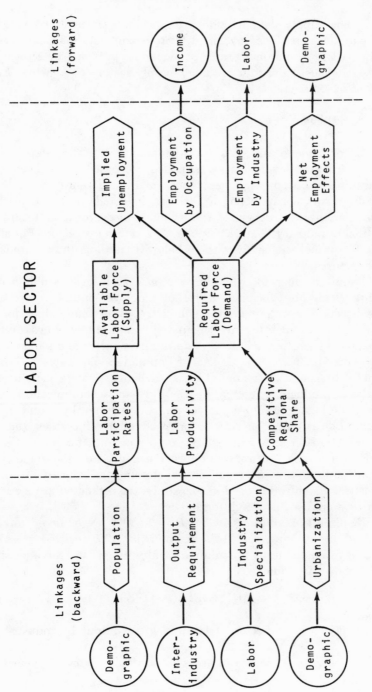

**Figure 9.** Components and sectoral linkages of the labor sector.

Labor supply data include area-specific participation rates and occupational profiles developed from census data [53] and rates of change in national labor force participation provided by Mincer [42, 58]. The calculations for the labor supply sector are shown as Equations 40–50 below.

**Labor Demand**

Labor demand estimates are calculated at both the reference and standard area levels. These estimates are based on realized gross output and output-employee ratios (at the reference region level) and industry occupational profiles (at the standard area level). The calculations also depend on shift-share disaggregation techniques utilizing industry control totals for the reference region. Equations 51–68 formalize the discussion below regarding these estimation procedures.

Required labor force at the reference area level is obtained by multiplying the diagonal vector of employee-output ratios times realized gross output for each industry. Base year values for the employee-output ratios are calculated using estimated gross output demanded as estimated in the interindustry sector and industry employment data from the Census of Population [53]. Table VII shows these ratios and their annual rates of change. Sectors showing high rates of productivity gain are agriculture and communications/utilities; the higher output-employee ratios are generally found in sectors subject either to substantial productivity increases (agriculture) or to services where labor inputs account for most of value added (finance, insurance and real estate). For those industries that required some aggregation to maintain consistency with the industrial classification used in this study, the estimated gross outputs in 1958 are used as weighting factors.

Realized gross outputs by industry in 1960 are assumed to equal gross output demanded. In years following 1960 realized gross output is estimated to be the minimum value among the three possibilities shown in Equation 35.

$$(OP)^t_i = MIN\,[(OPD)^t_i,\ (OPK)^t_i,\ (OPW)^t_i] \qquad (35)$$

where:

$(OP)^t_i$ = vector of realized gross output by industry $i$ in year $t$

$(OPD)^t_i$ = vector of gross output for industry $i$ in year $t$

Factor Input, Income and Resource Models    65

$(OPK)^t_i$ = capital-limited gross output for industry $i$ in year $t$

$(OPW)^t_i$ = water-limited gross output for industry $i$ in year $t$.

The use of $(OP)^t_i$ recognizes the various constraining influences on industry gross output that are present in the reference economy; its size is assumed to be responsive to a multiplicity of factors that influence the final demand vectors such as population change, income, investment, public spending and external demand. $(OPK)^t_i$ is assumed to be responsive to changes in capital stock levels and is

Table VII

Estimated Output-Employee and Annual
Rate of Change in Labor Productivity, Iowa, 1960

| Industry | Output per Employee[a] ($1000) | Change in Labor Productivity[b] (Annual Rate) |
|---|---|---|
| 1. Agriculture | 22.768 | 0.0530 |
| 2. Mining and construction | 17.894 | 0.0161 |
| 3. Food and kindred | 49.186 | 0.0240 |
| 4. Textile and apparel | 10.598 | 0.0262 |
| 5. Furniture, lumber and wood | 15.592 | 0.0348 |
| 6. Printing and publishing | 10.163 | 0.0270 |
| 7. Chemicals and allied | 61.194 | 0.0440 |
| 8. Machinery | 16.429 | 0.0081 |
| 9. Other and miscellaneous manufacturing | 19.300 | 0.0192 |
| 10. Transportation | 13.582 | 0.0400 |
| 11. Communication and utilities | 19.959 | 0.0607 |
| 12. Trade | 9.599 | 0.0140 |
| 13. Finance, insurance and real estate | 39.498 | 0.0100 |
| 14. Services | 6.889 | 0.0047 |
| 15. Public administration | 5.369 | 0.0000 |

[a]These estimates are based on gross output estimates from the input-output model in 1960 and employment by industry for the same year as reported in the Census of Population [54]. Small adjustments in employment were required to maintain consistency in industry classification with the input-output tables.

[b]These estimates are derived from U.S. labor productivity rates developed by Almon [1].

calculated in Equation 34 above. Its important determinants included the initial level of the capital stock by industry, the distribution and rate of investment in new capital and capital productivity. Given estimates of the three variables in Equation 35, labor demand by industry for the reference region is estimated by multiplying the inverse of $(OPE)^t{}_{ij}$ by the vector of realized gross output for the same year; this calculation is completed in Equation 63 below.

Standard area labor demand is estimated by using shift-share analysis [4, 5] as a disaggregating technique. The use of the shift-share model in this manner incorporates differences in subarea competitive position with respect to market access to essential factor inputs; these determine the competitive regional share in Figure 9. Shift-share analysis provides a means of examining changes in employment for small regions relative to a larger base region or nation. Three components of change in employment in a given industry $i$ for a given region $r$ are usually identified: first is the all-region, all-industry growth element for industry $i$ in region $r$ denoted by $(A)_{ir}$; second is the industry mix element in region $r$ denoted by $(B)_{ir}$; third is the regional share element for industry $i$ in region $r$ denoted by $(C)_{ir}$.

Defining employment as $(E)$ and letting the total change in employment between two points in time for industry $i$ of region $r$ be defined as $(D)_{ir}$, then

$$(D)_{ir} = (A)_{ir} + (B)_{ir} + (C)_{ir} \tag{36}$$

$$(A)_{ir} = \Sigma_i \, \Sigma_r \, \Delta(E)^\circ{}_{ir}{}^{-t} / \Sigma_i \Sigma_r (E)^\circ{}_{ir} \tag{37}$$

$$(B)_{ir} = [\Sigma_r \, \Delta(E)^\circ{}_{ir}{}^{-t} / \Sigma_r (E)^\circ{}_{ir}] - (A)_{ir} \tag{38}$$

$$(C)_{ir} = [\, \Delta(E)^\circ{}_{ir}{}^{-t} / (E)^\circ{}_{ir}] - (B)_{ir} \tag{39}$$

The regional share effect summarizes the competitive position of a particular industry within the region [36] and can be used for projection purposes only if suitable control totals on employment by industry are available for the reference region. From Equations 37 and 38 the growth and mix effects are estimated by aggregating over industries and regions. If total employment and employment by industry are known for the reference region in both the initial and terminal years, then it is possible to estimate employment change due to growth and mix effects if only initial year employment data are known for the subareas within the reference region. Further, if the subarea share effects (Equation 39) can be expected to remain constant over time, then small area employment by industry can be projected. These estimates are obtained by summing the products of the growth, mix and share coefficients times base year employment. The error encountered in the projection can be estimated by summing over regions by industry and comparing the sum to the reference area control totals. Though the share effect may not remain constant, the same procedure can be used if regional share coefficients are projected.

The projection of employment by industry will only by chance be exactly equal to the reference area control totals. Consistent disaggregation of the control totals by industry are obtained by multiplying these projected small region estimates by a correction factor equal to the control total divided by the sum of employment over all regions for the same industry.

Labor demand (or the required labor force) in standard areas is generated by using the shift-share technique to project and then disaggregate year by year with each annual iteration of the model. Control totals on employment demand by industry for the reference region are based on the product of current year realized gross output and the output per employee ratios. Growth and mix coefficients are developed from these totals and regional share effects are estimated by evaluating two sets of regression equations. The first set, shown as Equation 54 below, is estimated by using data from standard areas (1–99) that were predominantly rural. The second set, Equation 55 below, utilized data from standard areas (100–120) that were urban areas, urban places and cities (CDAs) with populations exceeding 10,000 in 1960. A full tabulation and discussion of these results may be found in Fullerton [16, p. 104 f.].

The regression equations were estimated to improve on the reliability of the regional share coefficients used in the disaggregation procedure. Though most variables generated within the model did not yield satisfactory statistical estimates, the best results for standards areas were obtained with the following independent variables: $X_1$ = primary employment proportion of total employment; $X_2$ = manufacturing employment proportion of total employment; $X_3$ = service employment proportion of total employment; $X_4$ = percentage change in population; $X_5$ = urbanization dummy variable equal to one if the area contained a CDA, zero if not; $X_6$ = stratification dummy variable equal to one if the area contained a population greater than 30,000, zero if not. Seven of the 15 industries yielded $R^2$ greater than 0.27; the range for $R^2$ is between 0.27 and 0.49 and the excluded industrial sectors are (1) and (3–9). In standard areas (100–120) the same independent variables are included with the following exceptions: $X_6$ is not used and $X_5$ is replaced by an SMSA dummy variable where the variable is equal to one if the CDA is also an SMSA and a zero if not. Eleven of the industries yielded $R^2$ exceeding 0.29, and the range of $R^2$ is between 0.29 and 0.57; the excluded industries are sectors (3), (4), (6) and (15).

The range of choice of potential independent variables used in estimating these equations is limited due to the manner in which the regressions are used. To adjust the regional share coefficients during each annual iteration the model must independently estimate the principal determinants of Equations 54 and 55. In the absence of this constraint, other independent variables could have been used and the statistical results would probably have been improved.

### Equations of the Labor Sector

Equations of the labor sector are given below. Unless otherwise noted, the definition and range of subscripts and superscripts are as follows:

$t = 0, 1, \ldots 20$; where $t$ refers to time in years and $t=0$ in 1960

$i = 1, 2, \ldots 15$; where $i$ refers to industry

$r = 1, 2, \ldots 120$; where $r$ refers to standard area

$s = 1, 2$; where $s$ refers to sex, s = 1 for males and s = 2 for females

$a = 1, 2, \ldots 16$; where $a$ refers to cohort $(0-4, 5-9, \ldots, 70-74, 75+)$

$m = 1, \ldots 6$; where $m$ refers to the labor group

$o = 1, 2, \ldots 8$; where $o$ refers to the occupational group.

### Labor Supply

$$(ALF)^t = \Sigma_r \, \Sigma_s \, \Sigma_m \, (ALF)^t_{rsm} \tag{40}$$

$$(ALF)^t_r = \Sigma_s \, \Sigma_m \, (ALF)^t_{rsm} \tag{41}$$

$$(ALF)^t_{rs1} = [(POP)^t_{rs3} * 0.2 + (POP)^t_{rs4} * 0.6] * (LFPR)^t_{rs1} \tag{42}$$

$$(ALF)^t_{rs2} = [(POP)^t_{rs4} * 0.4 + (POP)^t_{rs5} * 1.0] * (LFPR)^t_{rs2} \tag{43}$$

$$(ALF)^t_{rs3} = [\sum_{a=6}^{7} (POP)^t_{rsa}] * (LFPR)^t_{rs3} \tag{44}$$

$$(ALF)^t_{rs4} = [\sum_{a=8}^{9} (POP)^t_{rsa}] * (LFPR)^t_{rs4} \tag{45}$$

$$(ALF)^t_{rs5} = [\sum_{a=10}^{13} (POP)^t_{rsa}] * (LFPR)^t_{rs5} \qquad (46)$$

$$(ALF)^t_{rs6} = [\sum_{a=14}^{16} (POP)^t_{rsa}] * (LFPR)^t_{rs6} \qquad (47)$$

$$(LFPR)^t_{sm} = [(GLP)^o_{sm} ** t] * (LFPR)^o_{rsm} \qquad (48)$$

$$(ALF)^t_{rs} = \Sigma_m (ALF)^t_{rsm} \qquad (49)$$

$$(ALFOC)^t_{so} = (OCP)^o_{so} * (ALF)^t_s \qquad (50)$$

**Labor Demand**

$$(RLF)^t_{ir} = (RLF)^{t-1}_{ir} + (RLF)^{t-1}_{ir} * [(SGC)^t + (IMC)^t_i + (RSC)^t_{ir}]$$
$$(51)$$

$$(SGC)^t = [(RLF)^t / (RLF)^{t-1}] - 1 \qquad (52)$$

$$(IMC)^t_i = [(RLF)^t_i / (RLF)^{t-1}_i] - (SGC)^t \qquad (53)$$

$$(RSC)^t_{ir} = b_0 + b_1 (EP)^{t-1}_r + b_2 (EM)^{t-1}_r + b_3 (ES)^{t-1}_r + b_4 (PCP)^t_r$$
$$+ b_5 (LD)^o_r + b_6 (SC)^o_r$$
$$r = 1, 2, \ldots 99 \qquad (54)$$

$$(RSC)^t_{ir} = b_0 + b_1 (EP)^{t-1}_r + b_2 (EM)^{t-1}_r + b_3 (ES)^{t-1}_r + b_4 (PCP)^t_r$$
$$+ b_5 (SMSAD)^o_r$$
$$r = 100, 101, \ldots 120 \qquad (55)$$

$$(NEE)^t_r = [\Sigma_i (RLF)^t_{ir} - \Sigma_i (RLF)^{t-1}_{ir}] - [(SGC)^t * \Sigma_i (RLF)^{t-1}_{ir}]$$
$$(56)$$

$$(SRLF)^t_i = (OPE^{-1})^t_{ij} * (OP)^t_i \qquad (57)$$

$$\Sigma_r (RLF)^t_{ir} = (SRLF)^t_i \qquad (58)$$

$$(EP)^t_r = \sum_{i=1}^{2} (RLF)^t_{ir} / \sum_{i=1}^{15} (RLF)^t_{ir} \qquad (59)$$

$$(EM)^t_r = \sum_{i=3}^{9} (RLF)^t_{ir} / \sum_{i=1}^{15} (RLF)^t_{ir} \qquad (60)$$

$$(ES)^t_r = \sum_{i=10}^{15} (RLF)^t_{ir} / \sum_{i=1}^{15} (RLF)^t_{ir} \qquad (61)$$

$$(PCP)^t_r = [\Sigma_a \Sigma_s (POP)^t_{rsa} - \Sigma_a \Sigma_s (POP)^{t-1}_{rsa}] / \Sigma_a \Sigma_s (POP)^{t-1}_{rsa}$$
$$(62)$$

$$(OPE)^t_{ij} = [(GOE)_{ij} ** t] * (OPE)^o_i \qquad (63)$$

$$(SRLF)^t = \Sigma_i (SRLF)^t_i \qquad (64)$$

$$(IUR)^t = 1 - [(SRLF)^t / (ALF)^t] \tag{65}$$

$$(SRLFOC)^t_{io} = (OCPR)^o_{io} * (SRLF)^t_i \tag{66}$$

$$(SRLFOC)^t_o = \Sigma_i (SRLFOC)^t_{io} \tag{67}$$

$$(IUROC)^t_o = 1 - [(SRLFOC)^t_o / (ALFOC)^t_o] \tag{68}$$

where:

| | |
|---|---|
| $(ALF)^t_{rsm}$ | = available labor force in standard area $r$, of sex $s$, in age group $m$ in year $t$ |
| $(POP)^t_{rsa}$ | = population in standard area $r$, of sex $s$, cohort $a$ in year $t$ |
| $(LFPR)^t_{rsm}$ | = labor force participation rate in standard area $r$, of sex $s$, in age group $m$ in year $t$ |
| $(GLP)^o_{sm}$ | = diagonal matrix with nonzero elements equal to one plus the annual rate of change in labor force participation by sex $s$ and age group $m$ in the base year 1960 |
| $(ALFOC)^t_{so}$ | = labor force of sex $s$, occupation $o$, in year $t$ |
| $(OCP)^o_{so}$ | = diagonal matrix with nonzero elements equal to the proportion of individuals of sex $s$, in occupation $o$ in the base year 1960 |
| $(RLF)^t_{ir}$ | = required employment in industry $i$, standard area $r$ in year $t$ |
| $(SGC)^t$ | = growth coefficient in year $t$ |
| $(IMC)^t$ | = industry mix coefficient for industry $i$ in year $t$ |
| $(RSC)^t_{ir}$ | = regional share coefficient for industry $i$, standard area $r$, in year $t$ |
| $(EP)^t_r$ | = vector with elements representing primary employment as a proportion of total employment in standard area $r$ in year $t$ |
| $(EM)^t_r$ | = vector with elements representing manufacturing employment as a proportion of total employment in standard area $r$ in year $t$ |
| $(ES)^t_r$ | = vector with elements representing service employment as a proportion of total employment in standard area $r$ in year $t$ |
| $(PCP)^t_r$ | = vector with elements representing percentage change in total population between years $t-1$ and $t$ in standard area $r$, in year $t$ |
| $(LD)_r$ | = vector with elements representing a location dummy variable equal to one if standard |

area $r$ contained a CDA, a zero if not

| | |
|---|---|
| $(SC)_r$ | = vector with elements representing a stratification dummy variable equal to one if standard area $r$ contained a population in excess of 30,000, a zero if not |
| $(SMSAD)_r$ | = vector with elements representing a dummy variable equal to one if standard area $r$ contained an (SMSA) a zero if not |
| $(NEE)^t{}_r$ | = vector with elements representing the net employment effect in standard area $r$ in year $t$ |
| $(SRLF)^t{}_i$ | = vector of required employment in industry $i$ in year $t$ |
| $(OPE)^t{}_{ij}$ | = diagonal matrix with nonzero elements equal to the ratio of realized gross output from industry $i$ to employment in the same industry in year $t$ |
| $(OP)^t{}_i$ | = vector of realized gross output for industry $i$ in year $t$ |
| $(GOE)^o{}_{ij}$ | diagonal matrix with nonzero elements equal to one plus the annual rate of change in labor productivity per worker in industry $i$ in base year 1960 |
| $(IUR)^t$ | = implied unemployment rate for the reference region in year $t$ |
| $(SRLFOC)^t{}_{io}$ | = matrix with required employment profile by occupation $o$, industry $i$ for the reference region in year $t$ |
| $(OCPR)^o{}_{io}$ | = diagonal matrix with nonzero element equal to employment by occupation $o$ as a proportion of total employment in industry $i$ in the base year 1960 |
| $(IUROC)^t{}_o$ | = vector of implied unemployment rates by occupation $o$, within the reference region in year $t$. |

## INCOME SECTOR

This sector provides income estimates at both the reference and standard area levels. As indicated in Figure 10 income is directly related to value added estimates at the reference region level. State level control totals are then disaggregated to standard areas in a

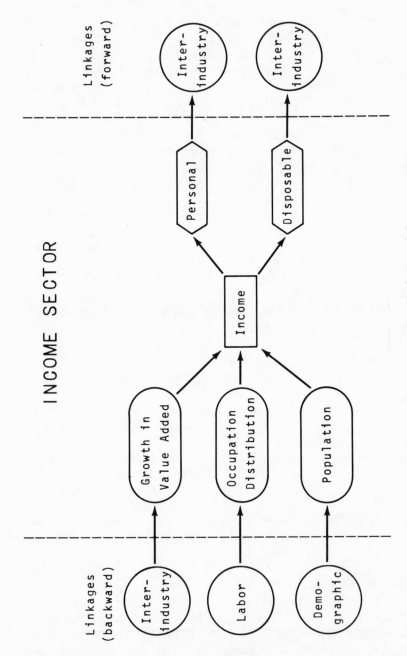

**Figure 10.**  Components and sectoral linkages of the income sector.

manner reflecting occupational specialization. Income estimates for standard areas provide useful status indicators of growth and development; at the reference region level income additionally influences the components of final demand in the interindustry sector.

## Reference Region Income

At the reference region changes in personal income are assumed to be directly related to changes in total value added. Prior to 1968, estimates of personal income from secondary sources [57, 59] are introduced directly into the model. In subsequent years personal income in year $t$ is calculated as the product of the ratio of total value added in year $t$ to total value added in year $t-1$ and personal income in $t-1$ (see Equation 69 below). Total value added is based on a summation of industry specific ratios of value added per dollar of gross output taken from the Iowa input-output model [30]. Annual estimates of value added by industry are the product of these value added-output ratios and realized gross output by industry for the same years. Personal income is then converted to disposable income by subtracting estimates of state and federal income taxes from personal income; both state and federal tax rate equations are developed for this purpose. These relationships are depicted in Figure 10 and specifically calculated in Equations 73 and 74 below.

## Standard Area Income

Standard area income estimates are disaggregated from the reference region. In each year a vector of income weights specific to occupation groups is estimated by multiplying the number of employed persons in each occupation group by a vector of income relatives. The vector of income relatives is based on the interrelation of average earnings of the major occupational groups used in this study, and estimates of employed persons by occupation group are obtained from the labor sector. The income relatives are standardized on the average earnings of professional-technical persons employed in the reference region.

The vector of income weights obtained above are subsequently used to estimate a set of proportions that express total personal income attributable to a particular occupational group. These proportions are used to distribute total personal income to occupations, and per capita figures are calculated by dividing by the number of employees in each group. Finally, estimates of standard area personal income are obtained by multiplying average income per employee in each occupation by the number of employees within each group in each standard area. Area personal income is estimated by summing over occupation by area; this sequence of calculations is shown below in Equations 77 through 84.

This procedure is an alternative to the use of industry average wage rates as a basis for estimating area income [12, 28]. The merit of this procedure rests on the extent to which occupational average earnings vary less among standard areas than average industrial wage rates. This, in turn, will depend on intraregional occupational mobility and the extent to which unions are able to bargain in selected industries for a wage package at least covering the reference region. Empirical tests of this relationship could be conducted but were outside the scope of this study.

### Income Sector Equations

Equations of the income sector are given below. Unless otherwise noted, the definition and range of subscripts and superscripts are as follows:

$t = 0, 1, \ldots 20$; where $t$ refers to time in years and $t=0$ in 1960

$i = 1, 2, \ldots 15$; where $i$ refers to industry

$r = 1, 2, \ldots 120$; where $r$ refers to standard area

$o = 1, 2, \ldots 8$; where $o$ refers to occupation category of the employed labor force.

### Reference Area Equations

$$(YP)^t = [(VA)/(VA)^{t-1}] * (YP)^{t-1} \tag{69}$$

$$(VA)^t = \sum_{i=1}^{15} (VA)^t_i \tag{70}$$

$$(VA)^t_i = (VAOP)^o_{ij} * (OP)^t_i \tag{71}$$

$$(YD)^t = (YP)^t - [(SYTX)^t + (FYTX)^t] \tag{72}$$

$$(SYTX)^t = A_1 + [(R_3 * t) * (YP)^t]$$
$$A_1 = 0.00756668$$
$$R_3 = 0.00339924 \tag{73}$$

$$(FYTX)^t = A_2 + [(R_4 * t) * (YP)^t]$$
$$A_2 = 0.07973841$$
$$R_4 = 0.00062084 \tag{74}$$

$$(YDPC)^t = (YD)^t / (TPOP)^t \tag{75}$$

$$(YPPC)^t = (YD)^t / (TPOP)^t \tag{76}$$

## Standard Area Equations

$$YWOC^t_o = (RLFOC)^t_o * (RYWOC)^o_o \tag{77}$$

$$(YW)^t_o = (YWOC)^t_o / \overset{8}{\underset{o=1}{\Sigma}} \; (YWOC)^t_o \tag{78}$$

$$(TYPOC)^t_o = (YW)^t_o * (YP)^t \tag{79}$$

$$(AVYPOC)^t_o = (TYPOC)^t_o / (RLFOC)_o \tag{80}$$

$$(RLFOC)^t_{ior} = (OCPR)^o_{io} * (RLF)^t_{ir} \tag{81}$$

$$(RLFOC)^t_{or} = \overset{15}{\underset{i=1}{\Sigma}} (RLFOC)^t_{ior} \tag{82}$$

$$(YPOC)^t_{or} = (RLF)^t_{or} * (AVYPOC)^t_o \tag{83}$$

$$(AYP)^t_r = \overset{8}{\underset{o=1}{\Sigma}} (YPOC)^t_{or} \tag{84}$$

where:

$(YP)^t$ = total personal income for the reference region in year $t$

$(VA)^t$ = total value added on production in the reference region in year $t$

$(VA)^t_i$ = value added in industry $i$ in year $t$

$(YD)^t$ = total disposable income for the reference region in year $t$

$(YDPC)^t$ = per capita disposable income for persons residing in the reference region in year $t$

$(SYTX)^t$ = state tax on personal income in year $t$

$(FYTX)^t$ = federal tax on personal income in year $t$

$(YPPC)^t$ = per capital personal income for persons residing in the reference region in year $t$

$(VAOP)^o_{ij}$ = diagonal matrix with nonzero elements to the ratio of value added per dollar of gross output from industry $i$ in the base year 1960

$(YWOC)^t_o$ = vector of income weights for occupation $o$ in year $t$

$(RYWOC)^o_o$ = vector of relative income weights associated with occupation $o$ in year $t$

$(YW)^t_o$ = vector expressing the proportion of personal income that will be allocated to occupation group $o$ in year $t$

$(TYPOC)^t_o$ = vector of total personal income for occupation group $o$ in year $t$

$(AVYPOC)^t_o$ = vector of average personal income for occupational group $o$ in year $t$

$(RLF)^t_{ior}$ = required employment in industry $i$ of occupational group $o$ in standard area $r$ in year $t$

$(YPOC)^t_{or}$ = personal income for occupation group $o$ in standard area $r$ in year $t$

$(AYP)^t_r$ = total personal income for standard area $r$ in year $t$

$(RLFOC)^t_{io}$ = required employment by occupation for industry $i$ in year $t$

$(OCPR)_{io}$ = diagonal matrix with nonzero elements expressing employment by occupation as a proportion of total employment in industry $i$ in the base year 1960.

## WATER SECTOR

There are two major purposes for including a water sector. First, this model provides estimates of implied water requirements over time for a variety of alternative subareas of the reference region. Second, this sector provides a means of tracing the impact of alternative levels of water availability on several important economic and demographic variables. When used in a projective context, these estimates assume a technology and policy structure reflected in base year data; therefore the model was designed to accommodate any set of anticipated water use rates that the analyst might wish to hypothesize. For base run purposes existing technological conditions are assumed. The flow chart shown in Figure 11 reflects these assumptions and is used in the description of the water sector's principal linkages as discussed below.

### Reference Area Water Requirement

Annual estimates of reference area water requirements by industry and domestic uses are calculated initially. The industrial water requirement is evaluated as the product of realized gross output by industry and a matrix of water disposition and use. Data on water

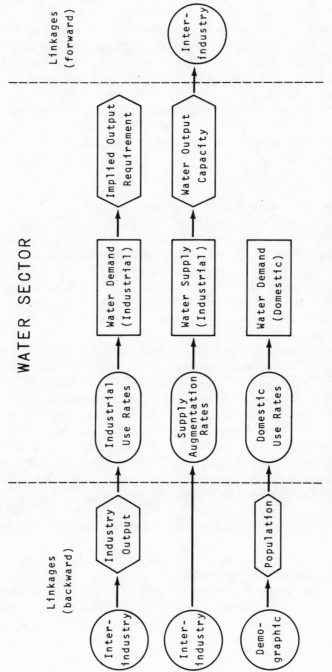

**Figure 11.** Components and sectoral linkages of the water sector.

use [25, 26, 27, 46, 55, 56] and studies by Barnard [7] and Martin and Carter [38] provided the basis for estimating the matrix of industrial water use; the 1960 estimates for Iowa are shown in Table VIII. In this table gross usage includes reuse, and consumption is the difference between water intake and discharge. Since the industrial use rates are based on 1960 input-output coefficients and output is annually estimated from the interindustry sector, the industrial water demand estimates similarly assume a constant relation among relative output prices and technologies influencing the efficiency of all inputs among industrial sectors. Equations 85 through 87 summarize the calculations involved in estimating water requirements by type of use and by industry.

### Standard Area Water Requirement

Industrial estimates of water use by standard area utilized the disaggregated employment estimates described in the labor sector above. The matrix of water intake, use, discharge and consumption is multiplied by industry estimates of realized gross output at the standard area level; since the reference region matrix is utilized it is assumed that water use ratios do not vary among subareas within the reference region. Industry employment by standard area is used as the proportioning factor in estimating standard area gross outputs and industrial water requirements. Unlike the studies of Barnard [7] and MacMillan [28] in which exogenous employment forecasts are used to spatially disaggregate water requirements, the dissaggregation techniques described in the labor sector allow for endogenously estimated and internally consistent forecasts of gross water use by standard area. Thus, the water use estimates will be sensitive to contractions or expansions of activity within other sectors of the entire model recognizing the unique standard area differences in competitive position and market access.

### Domestic Requirements and Water-limited Gross Output

Domestic water requirements are directly related to population levels as shown both in Figure 11 and Equation 90 below. The domestic component of total water requirements is estimated as the product of population and per capita use rates provided by the framework study of the Upper Mississippi River [52]. These estimates are of particular significance to CDA regions and assume that cost-use relationships implied in the per capita consumption figures

Table VIII

Estimated Water Intake, Use, Discharge and Consumption Coefficients, Iowa, 1960

| Industry | Gallons Per Dollar of Output | | | |
|---|---|---|---|---|
| | Water Intake | Gross Usage | Discharge | Consumption |
| 1. Agriculture | 11.70 | 11.70 | — | 11.70 |
| 2. Mining and construction | 25.69 | 56.68 | 18.00 | 7.69 |
| 3. Food and kindred | 17.31 | 27.34 | 16.06 | 1.25 |
| 4. Textile and apparel | 33.72 | 41.34 | 33.53 | 0.19 |
| 5. Furniture, lumber and wood | 82.03 | 107.59 | 74.99 | 7.04 |
| 6. Printing and publishing | 7.50 | 14.40 | 6.90 | 0.60 |
| 7. Chemicals and allied | 59.44 | 105.29 | 51.58 | 7.86 |
| 8. Machinery | 16.76 | 22.45 | 16.56 | 0.20 |
| 9. Other and miscellaneous manufacturing | 37.63 | 68.80 | 33.47 | 4.16 |
| 10. Transportation | 4.40 | 4.40 | 4.00 | 0.40 |
| 11. Communication and utilities | 807.34 | 868.26 | 804.79 | 2.55 |
| 12. Trade | 6.27 | 6.27 | 5.64 | 0.63 |
| 13. Finance, insurance and real estate | 1.63 | 1.63 | 1.45 | 0.18 |
| 14. Services | 25.51 | 25.51 | 22.94 | 2.57 |
| 15. Public administration | 35.16 | 35.16 | 31.61 | 3.55 |

will remain constant over the projection period. Though it seems likely that domestic use rates are relatively inelastic with respect to changes in municipal water charges, domestic use probably varies substantially by income level and net residential land density. However, very little work has been done on the relationship between income and domestic water use. Also, due to the absence of a land sector in the model, residential densities cannot be calculated within the model. As income and net residential densities will be correlated with city size, regionalizations that have a substantial variance in population among the largest cities will be affected most by the model's inability to account for these factors.

The water sector is also structured to allow for the impact on various demographic and economic variables of output levels restricted by limited water availability. In the base run water is assumed neither to restrict nor to enhance the economic and demographic variables, and estimates are calculated on the basis of Equation 85 below. In the experimental run the levels of water availability are reduced to 90% of the implied requirement given from the base run for selected industries, and the calculations for this restriction are summarized as Equations 86 and 87. It is assumed that water-limited output capacity would be responsive to public and private investments in water works, technological change in water use, institutional considerations including state and federal water policies, though these factors are not projected by the model. Though such institutional changes are less easily predicted, this restriction allows the model's user to study these interrelationships as a broad array of related changes that result from effective restraints on output due to water restrictions.

**Water Sector Equations**

Equations of the water sector are given below. Unless otherwise noted, the definition and range of subscripts and superscripts are are follows:

$t = 0, 1, \ldots . 20$; where $t$ refers to time in years and $t=0$ in 1960

$i = 1, 2, \ldots .$ where $i$ refers to the selling industry

$r = 1, 2, \ldots . 120$; where $r$ refers to standard area

$j = 1, 2, \ldots .$ where $j$ refers to the purchasing industry

$u = 1, 2, \ldots . 4$; where $u$ refers to water use of type.

*Reference Region Level*

$$(WR)^t_{iu} = (WOP)^o_{iu} * (OP)^t_i \tag{85}$$

$$(OPW)^t_i = (WOP^{-1})^o_{ij} * (SWS)^t_i \tag{86}$$

$$(SWS)^t_i = [(GWS)^o_{ij} ** t] * (SWS)^o_i \tag{87}$$

*Standard Area Level*

$$(WR)_{iur} = (WOP)^o_{iu} * (OP)^t_{ir} \tag{88}$$

$$(OP)^t_{ir} = [(RLF)^t_{ir} / (SRLF)^t_i] * (OP)^t_i \tag{89}$$

$$(DMWR)^t_r = (WUPC)^o * (POP)^t_r \tag{90}$$

where:

$(WR)^t_{iur}$    = matrix of water requirements by industry $i$ of use category $u$ in standard area $r$ in year $t$

$(WOP)^o_{iu}$    = diagonal matrix with nonzero elements equal to the ratio of water use by industry $i$ of use category $u$ in the base year 1960

$(OPW)^t_i$    = vector of water-limited gross output for industry $i$ in year $t$

$(SWS)^t_i$    = vector of water supply available (intake) to industry $i$ in year $t$

$(GWS)^o_{ij}$    = diagonal matrix with nonzero element equal to one plus the anticipated annual rate of growth in water supply to industry $i$ in the base year 1960

$(DMWR)^t_r$    = vector of water requirement for domestic use in standard area $r$ in year $t$

$(WUPC)^o$    = estimated per capital domestic water use rate in basic year 1960.

# 5

# APPLICATION OF
# THE SIMULATION MODEL

## INTRODUCTION

In this chapter the regional planning model is subjected to empirical tests designed to verify base run estimates and contrast projected variables under restrictive assumptions regarding future levels of water supply. The first section briefly compares model output during the historical phase 1961–70 with independent estimates from several studies of the region. Then trends in numerous development indices are summarized for the reference region over the base period; the latter includes both the historical and projective phase (1961–80) of the model. The next section contrasts the results obtained in a computer experiment where water supply is restricted and the base run in which unrestricted supply conditions were assumed. A final section provides a discussion of the results obtained when base run output was arrayed in different regional settings to determine the sensitivity of the impact and development indices to alternative spatial bases.

## MODEL VERIFICATION

As suggested by Forrester [13, p. 115], the effectiveness of a model depends on its structural breadth, the relevance of included variables, and numerical estimates of its parameters. Users of the information generated by the model determined its structural boundaries and most of the variables to be included; the spatial basis for disaggregation reflects the planning jurisdictions of the various user groups. All of these considerations must be rendered consistent with theoretical relationships suggested by the literature on regional economic modeling. Since the treatment of net migration in the demographic sector and the estimation of regional share coefficients in the labor sector are particularly unique features of this model, a more intensive statistical check was conducted on the population and employment estimates. Data insufficiencies preclude an exhaustive comparison of projected and actual magnitudes, but several independent studies provide information on these two sectors. These study estimates are compared to model output generated from a preliminary run covering the period 1960–70; reference region variables are utilized to assess the overall performance of the model.

### Demographic Sector

Preliminary census data for 1970 allowed us to assess model output in the demographic sector at several levels of spatial disaggregation. Table IX summarizes comparisons of total population estimates for the reference region, substate hydrologic basins (RBH) and economic river basins (RBE); similar tables for economic areas (EA) and urban counties (each with at least one city exceeding 10,000 in 1960) are found in Fullerton [16, pp. 127–128].

The model tends to overestimate reference region population and shows slightly improved results for smaller subregions. The reference region estimate is 2,815,045 in 1970, while the census figure is 24,109 lower, an underestimate of only 0.864%. Table IX also shows that the range of population differences in RBH and RBE was from 5,437 (low) to 95,229 (high); the largest percentage error (25.8%) occurs in RBH 5. All other differences are less than 12% for these larger substate regions, and generally improved results are obtained for the smaller areas. Among EA the range of differences in population is from 2,829 to 105,514 and the largest percentage difference is 21.4% in EA 8; only three differences among the 16 EAs are greater than 10.0%. Among CDA's the range of variability in population is from 128 tò 5,428; the percentage differences indicate a high of 14.3% (Fort Madison) and a low of 0.9% (Clinton).

## Labor Sector

Establishment data for· 1967 provided the first independent estimate used in checking employment calculations of the labor sector. Table X provides the data for a seven-industry comparison showing EAs with the highest and lowest percentage differences between the model and study-estimated total employment figures. (A full tabulation of all EAs is given in Fullerton [16, p. 130, Table 16].) In 1967 the model estimate for the entire reference region was 1,065,886 or about 4.5% below the independent study figure for all EAs in Iowa. As shown in Table X the range of percentage differences for total employment was from 0.9 to 24.6%; only two economic areas had employment estimates that differed by more than 10.0%.

Percentage differences between the model and independent study

### Table IX

**A Comparison of Model Population Estimates for the Reference Region RBH and RBE with Preliminary Census Estimates, 1970**

| Area | Model Estimate | Census Estimate | Difference Number | Per Cent[a] |
|---|---|---|---|---|
| Reference region | 2,812,045 | 2,787,936 | 24,109[b] | 0.9[b] |
| RBH 1 | 870,562 | 777,915 | 92,647[b] | 11.9[b] |
| 2 | 216,906 | 231,255 | 14,349 | 6.2 |
| 3 | 719,809 | 815,036 | 95,229 | 11.7 |
| 4 | 482,907 | 498,056 | 15,149 | 3.0 |
| 5 | 371,373 | 295,132 | 76,241[b] | 25.8[b] |
| 6 | 150,778 | 170,540 | 19,762 | 11.6 |
| RBE 1 | 941,775 | 858,327 | 56,448[b] | 6.5[b] |
| 2 | 111,061 | 116,498 | 5,437 | 4.7 |
| 3 | 811,697 | 827,435 | 15,738 | 1.9 |
| 4 | 422,370 | 430,727 | 8,357 | 1.9 |
| 5 | 472,222 | 494,643 | 22,421 | 4.5 |
| 6 | 53,110 | 60,306 | 7,196 | 11.9 |

[a]Difference between model estimate and preliminary census estimates was divided by the census estimate.

[b]In these areas model population estimates were greater than census estimates; hence, their difference is a negative number.

Table X

A Comparison of Model Employment Estimates by Industry for EAs 16 and 13
with Those Obtained in an Independent Study[a], 1967

| Industry[b] | EA 16 | | | EA 13 | | |
|---|---|---|---|---|---|---|
| | Model Estimate | Study Estimate | Difference in Per Cent | Model Estimate | Study Estimate | Difference in Per Cent |
| Agriculture | 4,632 | 4,632 | 0.0 | 13,272 | 13,475 | 1.5 |
| Mining and construction | 2,225 | 2,836 | 27.5 | 4,787 | 4,568 | 4.8 |
| Manufacturing | 12,902 | 18,420 | 42.7 | 10,296 | 10,236 | 0.5 |
| Transportation | 1,719 | 3,377 | 96.5 | 7,069 | 7,475 | 5.7 |
| Trade | 8,443 | 10,037 | 18.9 | 15,219 | 14,092 | 8.0 |
| Finance | 1,193 | 1,000 | 19.3 | 3,205 | 3,256 | 1.6 |
| Services | 11,747 | 13,107 | 11.6 | 17,191 | 18,551 | 7.9 |
| TOTAL | 42,861 | 53,409 | 24.6 | 71,039 | 71,653 | 0.9 |

[a]Study estimates were derived from unpublished data on employment by industry for 1967 as provided by Dr. Marvin Julius of the Department of Economics, Iowa State University.

[b]A complete designation of industries relative to the 15 Iowa industries is as follows: Agriculture (1), Mining and Construction (2), Manufacturing (3,4,5,6,7,8,9), Transportation (10,11), Trade (12), Finance (13), Services (14,15).

estimates were highest in the transportation sector. As indicated in Table X, EA 16 showed a difference of 96.5%, and among EAs 8 of the 16 regions had transportation employment estimates that differed by more than 30.0%. (In sectors other than transportation, percentage differences ranged from 0.0 to 38.6% with only 7 of the 96 comparisons exceeding 20.0%.) The two estimates of transportation employment may be noncomparable due to a different industrial classification used in the independent study; alternatively, the latter may not have attained a full translation of establishment employment data to the residentiary base comparable to the census data used in the model.

Additional employment comparisons show minor differences at the reference region level but rising percentage errors over the period 1965–70. Estimates provided by Maki [31] are shown in Table XI and indicate differences in reference region figures of 1.9 and 2.9% in these two years; the percentage differences for all RBE regions rise uniformly over this period. The model estimates are also generally lower than those calculated in this study, possibly due to the reduced rate of population growth both within the reference region and the United States in recent years. The estimates provided by Maki do not allow for feedback effects between the demographic and interindustry sectors that are specified in the simulation model.

The range of differences indicated in Tables X and XI provide a limited basis for assessing the accuracy of the labor sector of the overall model. Unlike census population figures, substantial errors in estimation may be present in these data. The 1967 data provide the most comparable information because they are based on a survey of employment at establishments and are subsequently converted to a residentiary base; their reliability depends on the accuracy of this conversion. The independent estimates in Table XI are projections based on historical trends and are calculated from a variant of the shift-share model. Given the possibility of estimating errors in these studies, the observed differences did not seem to warrant a major structural respecification of the labor sector.

## BASE RUN

Projective economic and demographic information was generated for the reference region and its subareas by allowing recursive iterations of the model to continue between 1970 and 1980. Though the model produces annual estimates, only five-year summaries are provided in the tables discussed below. Similarly, the spatial base

Table XI

A Comparison of Model Employment Estimates for the Reference Region and RBE with Those Obtained in an Independent Study[a], 1965 and 1970

| Area | 1965 | | | 1970 | | |
|---|---|---|---|---|---|---|
| | Model Estimate | Study Estimate | Difference in Per Cent | Model Estimate | Study Estimate | Difference in Per Cent |
| Reference region | 1,038,356 | 1,057,876 | 1.9 | 1,093,138 | 1,124,557 | 2.9 |
| RBE 1 | 342,698 | 324,036 | 5.7 | 378,374 | 342,050 | 10.6 |
| 2 | 42,410 | 51,083 | 20.4 | 42,809 | 56,524 | 32.0 |
| 3 | 301,420 | 315,669 | 4.7 | 314,193 | 337,980 | 7.6 |
| 4 | 146,794 | 161,925 | 10.3 | 150,993 | 178,030 | 17.9 |
| 5 | 181,014 | 181,249 | 0.1 | 182,855 | 186,265 | 1.9 |
| 6 | 24,020 | 23,915 | 0.4 | 23,918 | 23,710 | 0.9 |

[a]Study estimates were derived from tabular materials presented in Iowa economic trends: employment output income and related data [31]. Unpublished mimeographed paper prepared for the Iowa Office of Planning and Programming in 1968.

for the discussion includes only reference region variables, though the model generates information for 120 standard areas (and their various spatial combinations) over this 20-year period. Two problems were revealed by initial runs beyond the historical phase.

(1) Disparate growth rates in labor supply and demand were estimated after 1970. Labor demand rose at such a rapid rate that after 1975 the model generated negative unemployment. Low rates of projected labor force participation and/or worker productivity could cause this condition but both the initial rates and rates of change in the base data are obtained from reliable information [1, 8, 30, 42, 53]. Another possible cause was the rapid yearly changes in investment demand and disposable income per capita after 1970. Maki, *et al.* [37] suggests that financial and technical constraints probably provide an upper limit on capital stock expansion in any given year; in addition the model does not recognize constraints on capital availability to reference region firms that are either internal or external to the state. For these reasons an upper bound was placed on investment demand (Equation 30 in the capital sector) establishing a maximum rate of growth per year of 15% including depreciation. This constraint also dampened growth rates in per capital disposable income since the latter is indirectly dependent on investment demand through changes in realized output and capital accumulation.

(2) In the demographic sector, net migration rates tended to underestimate positive inmigration for age cohorts (4–7) in standard areas with major educational institutions. As a result the population growth rates for these areas were below estimates provided by preliminary census data. To improve the model's predictive ability, net migration rates for these cohorts were multiplied by an adjustment factor for standard areas 103 and 112. This factor assumed that the resident student population would maintain itself and inmigrating students would grow at a rate of 6% over the period 1960–80. The tables below reflect these two modifications resulting from an analysis of the model's initial run beyond the historical phase.

**Demographic Sector**

Base run projections for the demographic sector are consistent with mild overall rates of population increase and a rising average age for both males and females in the reference region. As shown in Table XII the total population is estimated to increase from 2,791,229

Table XII

Summary of Reference Region Population for the Years 1965, 1970, 1975 and 1980

| Age Group | 1965 Males | 1965 Females | 1970 Males | 1970 Females | 1975 Males | 1975 Females | 1980 Males | 1980 Females |
|---|---|---|---|---|---|---|---|---|
| 0-4 | 135,172 | 130,750 | 116,631 | 112,981 | 140,821 | 136,181 | 129,240 | 125,412 |
| 5-9 | 141,548 | 136,214 | 125,141 | 120,911 | 115,843 | 111,850 | 125,860 | 121,345 |
| 10-14 | 137,111 | 131,530 | 131,548 | 126,737 | 121,184 | 116,918 | 120,517 | 116,070 |
| 15-19 | 119,847 | 117,590 | 126,042 | 123,202 | 123,180 | 120,682 | 119,146 | 116,732 |
| 20-24 | 87,195 | 93,633 | 99,050 | 104,955 | 105,031 | 111,328 | 106,638 | 113,279 |
| 25-29 | 68,038 | 80,288 | 75,358 | 91,822 | 83,718 | 104,251 | 89,290 | 114,399 |
| 30-34 | 69,927 | 74,510 | 67,850 | 77,301 | 72,971 | 86,331 | 79,567 | 96,583 |
| 35-39 | 76,061 | 78,525 | 69,241 | 74,653 | 68,976 | 78,428 | 73,658 | 87,246 |
| 40-44 | 78,099 | 81,274 | 72,228 | 76,321 | 68,420 | 75,189 | 69,328 | 79,830 |
| 45-49 | 76,016 | 79,952 | 72,393 | 77,453 | 68,016 | 74,639 | 65,887 | 75,275 |
| 50-54 | 72,436 | 75,335 | 70,873 | 75,493 | 67,846 | 73,630 | 64,905 | 72,287 |
| 55-59 | 65,960 | 69,804 | 65,492 | 70,951 | 63,890 | 70,685 | 61,455 | 69,429 |
| 60-64 | 61,352 | 67,044 | 61,921 | 68,916 | 61,647 | 70,023 | 60,464 | 69,999 |
| 65-69 | 55,257 | 62,675 | 56,664 | 65,512 | 57,560 | 67,820 | 57,702 | 69,233 |
| 70-74 | 46,763 | 56,429 | 49,979 | 62,263 | 52,461 | 66,931 | 54,137 | 70,520 |
| 75+ | 69,561 | 95,409 | 91,445 | 130,915 | 112,671 | 169,899 | 133,215 | 212,086 |
| TOTAL[a] | 1,360,308 | 1,430,921 | 1,351,818 | 1,460,341 | 1,384,198 | 1,534,745 | 1,410,969 | 1,609,668 |

[a]Total may be more or less than the sum over all cohorts because of rounding error.

(1965) to 3,020,637 (1980), representing an average annual percentage increase of 0.5%. Among 32 age-sex cohorts the population of females in age groups 25–29 and 75+ showed the largest increase of 42.5 and 122.3% over the period 1965–80. The female population declined in 7 of 16 age groups and the male population declined in 8 age groups; uniform declines for both sexes occurred in the first four cohorts and increases were common in the last three age groups. The declines in population expected from post-1960 fertility experience were accentuated by negative net migration in the younger cohorts.

Reference region births and deaths tend to vary in proportion to their respective dependent cohorts due to the assumed constancy of birth and death rates over the simulation period; Tables XIII and XIV summarize reference region birth and death estimates by age-sex cohort. Births, by age of mother, varied in proportion to the female population in cohorts 4–9 after 1970 since 1968 fertility rates are held constant over the period. Similarly, the estimates of deaths in Table XIV are proportional to cohort population levels due to the assumed constancy of 1968 death rates. Population increases due to births may be overestimated if birth rates continue to fall and marriages are deferred; the relative stability in death rates probably results in more accurate mortality estimates.

Table XV shows the reversal of negative net migration rates for the total reference region population and substantial differences in migrational behavior among age cohorts. Through 1970 negative net migration for both sexes is experienced and this trend is reversed for females in 1975 and males in 1980. In part this result is due to the aging of the population into the last three age cohorts where net migration rates were held constant over the simulation period. The highest levels of negative net migration are shown for both males and females in age cohorts 0–4 and 20–24 with substantial numerical declines in most labor active groups. Positive net migration for females is estimated for age cohort 25–29 in both 1970 and 1975; similar behavior is shown for the 35–39 and 40–44 age groups in 1980. Net migration for males was uniformly negative for all age groups under 60 years until 1980 when positive net flows are estimated for age groups 35–39, 40–44 and 50–54.

## Interindustry Sector

Table XVI shows the five major categories of final demand, gross output demanded, and realized gross output by industry for 1965

### Table XIII

**Summary of Reference Region Births by Sex
and by Age of Mother in 1965 and 1980**

| Age Group of Mothers | 1965 | | 1980 | |
|---|---|---|---|---|
| | *Males* | *Females* | *Males* | *Females* |
| 15-19 | 3,248 | 3,123 | 2,970 | 2,838 |
| 20-24 | 9,300 | 8,942 | 9,664 | 9,137 |
| 25-29 | 6,785 | 6,519 | 7,554 | 7,328 |
| 30-34 | 3,926 | 3,778 | 4,064 | 4,185 |
| 35-39 | 2,102 | 2,026 | 1,901 | 2,026 |
| 40-44 | 622 | 600 | 516 | 543 |
| | | | | |
| TOTAL | 25,983 | 24,988 | 26,669 | 26,057 |

### Table XIV

**Summary of Reference Region Deaths by Age and Sex in 1965 and 1980**

| Age Group | 1965 | | 1980 | |
|---|---|---|---|---|
| | *Males* | *Females* | *Males* | *Females* |
| 0-4 | 986 | 675 | 918 | 633 |
| 5-9 | 77 | 46 | 63 | 38 |
| 10-14 | 51 | 10 | 47 | 10 |
| 15-19 | 151 | 25 | 138 | 26 |
| 20-24 | 152 | 11 | 162 | 20 |
| 25-29 | 128 | 49 | 152 | 65 |
| 30-34 | 126 | 53 | 133 | 67 |
| 35-39 | 185 | 137 | 162 | 139 |
| 40-44 | 274 | 174 | 230 | 158 |
| 45-49 | 499 | 238 | 417 | 215 |
| 50-54 | 707 | 335 | 622 | 313 |
| 55-59 | 1,019 | 471 | 960 | 468 |
| 60-64 | 1,471 | 762 | 1,444 | 788 |
| 65-69 | 1,977 | 1,201 | 2,029 | 1,304 |
| 70-74 | 2,356 | 1,544 | 2,639 | 1.811 |
| 75+ | 6,585 | 6,920 | 11,458 | 12,929 |
| | | | | |
| TOTAL | 16,744 | 12,651 | 21,574 | 18,984 |

Table XV

Summary of Reference Region Net Migration in 1965, 1970, 1975 and 1980

| Age Group | 1965 Males | 1965 Females | 1970 Males | 1970 Females | 1975 Males | 1975 Females | 1980 Males | 1980 Females |
|---|---|---|---|---|---|---|---|---|
| 0-4 | -2,674 | -2,570 | -1,739 | -1,813 | -1,248 | -1,450 | - 603 | - 952 |
| 5-9 | -1,722 | -1,655 | -1,056 | -1,078 | - 739 | - 829 | - 345 | - 515 |
| 10-14 | -1,221 | -1,177 | - 775 | - 777 | - 591 | - 632 | - 378 | - 470 |
| 15-19 | -1,685 | -1,200 | -1,599 | -1,007 | -1,332 | - 677 | -1,011 | - 334 |
| 20-24 | -4,185 | -2,843 | -4,380 | -2,712 | -3,988 | -2,129 | -3,265 | -1,347 |
| 25-29 | -2,201 | - 341 | -2,151 | 364 | -2,023 | 1,268 | -1,694 | 2,404 |
| 30-34 | - 840 | -1,392 | - 751 | -1,370 | - 729 | -1,466 | - 673 | -1,588 |
| 35-39 | - 382 | - 455 | - 185 | - 258 | - 26 | 25 | 166 | 265 |
| 40-44 | - 421 | - 327 | - 259 | - 201 | - 109 | - 45 | 33 | 148 |
| 45-49 | - 488 | - 312 | - 423 | - 241 | - 346 | - 144 | - 278 | - 27 |
| 50-54 | - 201 | - 347 | - 145 | - 332 | - 68 | - 298 | 18 | - 252 |
| 55-59 | - 290 | - 347 | - 278 | - 348 | - 255 | - 339 | - 222 | - 316 |
| 60-64 | 706 | 603 | 775 | 669 | 840 | 727 | 909 | 790 |
| 65-69 | 1,001 | 881 | 1,164 | 1,059 | 1,285 | 1,189 | 1,390 | 1,298 |
| 70-74 | 1,296 | 1,462 | 1,640 | 1,934 | 1,920 | 2,313 | 2,134 | 2,610 |
| 75+ | 1,263 | 1,926 | 2,248 | 3,431 | 3,448 | 5,371 | 4,728 | 7,626 |
| TOTAL | -12,044 | -8,094 | -7,914 | -2,680 | -3,961 | 2,834 | 909 | 9,340 |

Table XVI

Summary of Interindustry Sector Data for the Years 1965 and 1980 (in thousands of dollars)

| Industry | 1965 Final Demands | | | | | | 1965 Output | |
|---|---|---|---|---|---|---|---|---|
| | PCE | CA | FGE | SLGE | EX | Total^a | OPD | OP |
| Agriculture | 78,309 | — | 4,852 | - 886 | 2,149,823 | 2,232,096 | 4,482,383 | 4,482,383 |
| Mining and construction | 4,653 | 53,885 | 4,376 | 301,320 | 28,259 | 392,493 | 646,143 | 582,106 |
| Food and kindred | 735,880 | — | 173 | 7,311 | 1,449,178 | 2,192,540 | 4,671,871 | 4,671,871 |
| Textile and apparel | 229,293 | 150 | 28 | 376 | 134 | 229,980 | 289,490 | 289,490 |
| Furniture, lumber and wood | 44,889 | 250 | 28 | 3,882 | 1,149 | 50,197 | 81,720 | 81,720 |
| Printing and publishing | 41,323 | 35 | 99 | 4,643 | 4,973 | 51,072 | 77,701 | 77,701 |
| Chemicals and allied | 63,420 | 39 | 283 | 6,945 | 74,919 | 145,607 | 219,030 | 219,030 |
| Machinery | 84,036 | 500,082 | 1,300 | 6,772 | 213,322 | 805,512 | 1,206,752 | 1,171,351 |
| Other and miscellaneous manufacturing | 523,676 | 108,330 | 951 | 15,817 | 82,583 | 731,357 | 1,112,179 | 1,078,775 |
| Transportation | 145,898 | 30,554 | 1,518 | 9,440 | 38,505 | 225,914 | 313,932 | 313,932 |
| Communication and utilities | 199,298 | 5,460 | 2,356 | 14,796 | 27,329 | 249,239 | 361,653 | 361,653 |
| Trade | 1,017,893 | 62,837 | 991 | 4,913 | 280,696 | 1,367,328 | 1,810,814 | 1,810,814 |
| Finance, insurance and real estate | 886,113 | 1,816 | 117 | 10,269 | 4,850 | 903,164 | 1,278,733 | 1,264,725 |
| Services | 646,575 | — | 3,525 | 19,594 | 209,844 | 879,536 | 1,393,318 | 1,393,318 |
| Public administration | 16,075 | — | 201 | 1,927 | 64,166 | 82,368 | 133,370 | 133,370 |
| TOTAL | 4,717,331 | 763,438 | 20,798 | 407,119 | 4,629,730 | 10,538,416 | 13,641,529 | 13,494,679 |

^aTotal final demand may be slightly more or less than the sum over individual categories because of rounding error.

**Table XVI, continued**

| Industry | PCE | CA | 1980 Final Demands | | | | 1980 Output | |
| --- | --- | --- | --- | --- | --- | --- | --- | --- |
| | | | FGE | SLGE | EX | Total[a] | OPD | OP |
| Agriculture | 95,295 | — | 12,328 | - 1,550 | 2,988,399 | 3,094,469 | 6,214,158 | 6,214,158 |
| Mining and construction | 6,440 | 108,851 | 11,119 | 527,145 | 48,034 | 701,589 | 1,154,991 | 1,097,156 |
| Food and kindred | 873,252 | — | 439 | 12,791 | 2,379,078 | 3,265,560 | 6,958,265 | 6,958,265 |
| Textile and apparel | 287,466 | 189 | 72 | 657 | 233 | 288,618 | 363,301 | 363,301 |
| Furniture, lumber and wood | 61,904 | 353 | 71 | 6,792 | 1,902 | 71,021 | 115,621 | 115,621 |
| Printing and publishing | 61,981 | 54 | 251 | 8,123 | 8,726 | 79,135 | 120,394 | 120,394 |
| Chemicals and allied | 104,965 | 67 | 720 | 12,151 | 135,119 | 253,021 | 380,607 | 380,607 |
| Machinery | 120,754 | 1,343,439 | 3,302 | 11,847 | 460,893 | 1,940,235 | 2,906,701 | 2,906,701 |
| Other and miscellaneous manufacturing | 728,208 | 120,167 | 2,417 | 27,672 | 142,007 | 1,020,470 | 1,551,835 | 1,551,835 |
| Transportation | 205,771 | 81,182 | 3,857 | 16,515 | 70,554 | 377,879 | 525,103 | 512,929 |
| Communication and utilities | 275,054 | 99,173 | 5,986 | 25,885 | 58,128 | 373,245 | 541,589 | 541,589 |
| Trade | 1,347,817 | 124,951 | 2,519 | 8,595 | 483,378 | 1,931,197 | 2,557,572 | 2,557,572 |
| Finance, insurance and real estate | 1,698,630 | 226,082 | 298 | 17,965 | 9,841 | 1,730,054 | 2,449,474 | 2,449,474 |
| Services | 1,097,110 | 304,007 | 8,955 | 34,278 | 387,846 | 1,528,186 | 2,420,876 | 2,420,876 |
| Public administration | 23,247 | 39,504 | 511 | 3,370 | 152,693 | 179,821 | 291,163 | 291,163 |
| TOTAL | 6,987,894 | 2,448,019 | 52,845 | 712,236 | 7,326,831 | 17,527,825 | 28,551,650 | 28,481,641 |

aTotal final demand may be slightly more or less than the sum over individual categories because of rounding error.

and 1980. (The estimates for 1970 and 1975 are consistent with trends discussed below and are tabulated in Fullerton [16, p. 151 f.].)

The least predictable components of final demand are personal consumption expenditure (PCE) and purchases of goods and services by the federal government (FGE). Although population growth influences both PCE and FGE, their growth rates substantially exceed rates of increase in the demographic sector. Between 1965 and 1980 PCE and FGE increased by 48 and 154% respectively, while population rose by slightly more than 8%. This is explained by their respective dependence on per capita income generated within the model and a time trend in the rate of increase in FGE incorporated in the base run. Changes in capital accumulation (CA) and net exports (EX) were more stable because they are more dependent on trends determined outside of the model. As noted above, CA is constrained by an upper bound and the growth in EX was based on the assumption that each industry in the reference region would maintain a constant market share in export sales. CA, FGE and EX were included in the model in such a manner that several alternative assumptions regarding their magnitudes could be incorporated; base run levels and rates of change in these sectors reflect historical conditions.

A comparison of output demanded (OPD) and realized output (OP) for all industries in 1965 indicated that demand was an effective constraint on gross output in 11 of 15 industries including agriculture and food/kindred products (see the last two columns of Table XVI). Where OP is less than OPD in the base run, capital capacity is the limiting factor). By 1980 demand was limiting gross output in 13 of the 15 industries, and capital constraints occurred in the mining and construction and transportation sectors. Estimated trends in realized gross output show increases for all industries between 1965 and 1980. The largest percentage increases were in the construction, machinery, and finance, insurance and real estate industries, which experienced increases of 88, 148 and 93% respectively.

**Capital Sector**

The principal variables in the capital sector showed uniformly positive increases over the period 1965–80. Positive changes in the capital stock (KAP) occurred for all industries; the estimated total stock rose from 13.5 billion dollars (1965) to 21.5 billion dollars (1980), representing a 59% increase. Capital capacity (OPK) tended to vary in proportion to the capital stock since incremental output-capital ratios were assumed to remain constant over the period. Changes in investment (IVST) by industry were also uniformly positive; total investment increased from 1.2 billion dollars (1965) to 2.0 billion dollars (1980), or an increase of 60%.

## Water Sector

Table XVII provides point estimates of water intake, use, consumption and discharge for 1965 and the terminal year of the simulation. The use quantities vary directly with changes in realized gross output since the water output coefficients are assumed to remain constant over the period; this assumption may overstate water use to the extent that constant coefficients do not reflect water saving technologies or institutional practices conducive to reduced use levels. Domestic use varies in proportion to changes in total population since use rates per capita are utilized in the model. These rates were based on estimates used in the framework study of the upper Mississippi River basin and are also assumed to hold constant over the projection period.

## Labor Sector

Labor supplies projected by the model reflect both aging and a heavier reliance on female labor; Table XVIII shows estimated labor supply by age-sex cohorts for 1965 and 1980. The percentage of workers found in age groups 18–24 and 25–34 increased from 13.5 and 14.3% in 1965 to 17.2 and 18.9% in 1980. Between 1965 and 1980 the percentage of employment attributable to female workers rose from 27.8 to 38.4%. In labor age groups 18–24 females contributed 58.9% of the 34,134 increase in total labor supplies from 1965–80; similarly in age groups 25–34 females were 61.4% of the projected increase of 45,429. This result is consistent with the "quasi" full employment condition projected by the model because in tight labor markets female labor force participation tends to increase. Further evidence of this phenomenon is provided by the projection of positive net migration for females in the 25–29 age group since net migration rates are directly related to independent variables reflecting employment opportunities within the reference region.

Labor demand generally increased at rates exceeding population growth and showed a substantial variation among industries. Table XIX shows that total labor demand is estimated to increase from

Table XVII

Summary of Water Intake, Use, Consumptive Use and Discharge
for the Reference Region in the Years 1965 and 1980 (millions of gallons)

| Industry | 1965 | | | | 1980 | | | |
|---|---|---|---|---|---|---|---|---|
| | Intake | Use | Consumption | Discharge | Intake | Use | Consumption | Discharge |
| Agriculture | 52,443.9 | 52,443.9 | 52,443.9 | — | 72,705.6 | 72,705.6 | 72,705.6 | — |
| Mining and construction | 14,995.0 | 32,994.9 | 4,477.6 | 10,477.4 | 28,188.1 | 62,189.0 | 8,439.3 | 19,748.8 |
| Food and kindred | 80,870.1 | 127,729.0 | 5,840.8 | 75,029.3 | 120,447.6 | 190,239.0 | 11,202.8 | 109,244.8 |
| Textile and apparel | 9,762.6 | 11,967.5 | 60.9 | 9,701.7 | 12,250.5 | 15,018.9 | 76.3 | 12,174.2 |
| Furniture, lumber and wood | 6,703.5 | 8,792.3 | 575.3 | 6,128.2 | 9,484.4 | 12,439.7 | 813.9 | 8,670.5 |
| Printing and publishing | 583.8 | 1,118.9 | 47.7 | 536.1 | 903.0 | 1,733.7 | 72.2 | 830.8 |
| Chemicals and allied | 13,019.1 | 23,061.7 | 1,728.1 | 11,291.0 | 22,623.3 | 40,074.1 | 3,003.0 | 19,620.3 |
| Machinery | 19,642.0 | 26,293.3 | 239.6 | 19,402.4 | 48,719.2 | 65,246.7 | 569.7 | 48,149.5 |
| Other and miscellaneous manufacturing | 40,590.0 | 74,221.9 | 3,799.5 | 36,790.5 | 58,389.3 | 106,769.4 | 5,465.6 | 52,923.7 |
| Transportation | 1,381.3 | 1,381.3 | 126.6 | 1,254.7 | 2,256.9 | 2,256.9 | 205.2 | 2,051.7 |
| Communication and utilities | 291,975.5 | 314,009.9 | 3,475.5 | 288,500.0 | 437,244.3 | 470,241.7 | 5,204.7 | 432,040.6 |
| Trade | 11,353.8 | 11,353.8 | 1,140.8 | 10,213.0 | 16,036.0 | 16,036.0 | 1,611.3 | 14,424.7 |
| Finance, insurance and real estate | 2,061.5 | 2,061.5 | 227.7 | 1,833.8 | 3,992.6 | 3,992.6 | 440.9 | 3,551.7 |
| Services | 35,543.5 | 35,543.5 | 3,566.8 | 31,976.7 | 61,756.5 | 61,756.5 | 6,197.4 | 55,559.1 |
| Public administration | 6,689.3 | 4,689.3 | 473.5 | 4,215.8 | 10,237.3 | 10,237.3 | 1,033.6 | 9,204.7 |
| TOTAL | 585,574.9 | 726,662.7 | 78,224.3 | 507,350.6 | 905,234.6 | 1,130,937.1 | 117,041.5 | 788,195.1 |

## Table XVIII

### Summary of Labor Supply for the Reference Region in the Years 1965 and 1980

| Age Group | 1965 | | 1980 | |
|---|---|---|---|---|
| | Males | Females | Males | Females |
| 14-17 | 39,466 | 18,781 | 36,052 | 18,130 |
| 18-24 | 104,588 | 63,283 | 118,703 | 83,302 |
| 25-34 | 129,237 | 47,744 | 146,770 | 75,640 |
| 35-44 | 149,487 | 63,966 | 139,149 | 80,121 |
| 45-65 | 252,958 | 126,242 | 228,763 | 152,921 |
| 65+ | 50,644 | 24,893 | 53,349 | 41,390 |
| TOTAL[a] | 726,380 | 344,909 | 722,786 | 451,504 |

[a]Total may be more or less than the sum over all groups because of rounding error.

## Table XIX

### Summary of Labor Demand for the Reference Region in the Years 1965 and 1980

| Industry | 1965 | 1980 |
|---|---|---|
| Agriculture | 180,916 | 115,592 |
| Mining and construction | 62,881 | 93,269 |
| Food and kindred | 56,353 | 58,807 |
| Textile and apparel | 4,458 | 3,795 |
| Furniture, lumber and wood | 7,131 | 6,031 |
| Printing and publishing | 18,334 | 19,049 |
| Chemicals and allied | 4,918 | 4,480 |
| Machinery | 75,532 | 166,069 |
| Other and miscellaneous manufacturing | 50,564 | 54,685 |
| Transportation | 36,266 | 32,902 |
| Communication and utilities | 21,526 | 13,319 |
| Trade | 214,473 | 245,900 |
| Finance, insurance and real estate | 44,859 | 74,835 |
| Services | 218,534 | 353,914 |
| Public administration | 41,608 | 90,835 |
| TOTAL[a] | 1,038,356 | 1,333,482 |

[a]Total may be more or less than the sum over industries because of rounding error.

1,038,356 (1965) to 1,333,482 workers (1980) representing an annual average increase of 1.6%. The range of variation by industry over this period was −65,324 in agriculture to +135,380 in the service sector, a result consistent with labor productivity trends in the reference region economy. Since labor demand was directly related to realized gross output by industry and projected worker productivity ratios, errors in the latter will influence the estimates shown in Table XIX.

**Income Sector**

Table XX shows estimated trends in personal income per capita, per employee and per employee by occupation group in 1965 and 1980. The occupational groups comprising professional and technical workers, farmers and managers, sales workers, craftsmen and operatives were above the average income per employee for all projected years while the remaining groups were below average. Per capita income was projected to increase to $4,855 by 1980, or 77% above the 1965 level.

**EXPERIMENTAL RUN**

After completion of the base run, one of several possible experi-

Table XX

Summary of Average Personal Income Per Capita,
Per Employee and Per Employee by Occupation Group for Residents
of the Reference Region in the Years 1965 and 1980

| Receiving Unit | 1965 | 1980 |
|---|---|---|
| Per person | 2,733 | 4,855 |
| Per employee | 7,347 | 10,997 |
| Occupation: | | |
|   Professional and technical | 9,487 | 14,234 |
|   Farmers and managers | 8,427 | 12,643 |
|   Clerical workers | 5,987 | 8,848 |
|   Sales workers | 7,422 | 11,134 |
|   Craftsmen and foremen | 9,258 | 13,890 |
|   Operatives | 7,698 | 11,550 |
|   Service workers | 3,350 | 5,207 |
|   Laborers | 4,767 | 7,152 |

ments with the model was conducted. The experiment discussed here assumed that the water supply made available to manufacturing industries between 1970 and 1980 would expand at a rate that would result in an implied intake requirement in 1980 10% below the level indicated in the unconstrained base run. No other restrictions were imposed on this experiment, so changes in the direction and magnitude of the demographic and economic variables are due solely to levels of water availability that are below requirements implied by the base run simulation. Restrictions of this nature would be primarily due to natural phenomena, inadequate water resource development, or some combination of these causes. The 10% restriction is not an expected condition in the reference region but rather provides a sufficient shock so that the salient impacts on the model's estimates can be clearly distinguished.

The water restriction is intended to emphasize the type of resource policy decisions that must be made under limited supply conditions. An objective of the experiment is to identify clearly the industries and sectors that bear the principal direct effects of water restrictions. This information should facilitate the identification of economic sectors with a substantial stake in water development policy within the reference region in addition to providing aggregate estimates of economic losses due to limited water supplies. As in the base run, substantial spatial and temporal detail is provided by the model, but to facilitate the discussion only reference region variables are analyzed. Comparisons to base levels are emphasized; detailed estimates are provided in Fullerton [16, Appendix F].

**Demographic Sector**

Water restrictions constrain gross industry output, reduce the demand for labor and therefore influence net migration rates in the labor active cohorts 4–12. Direct employment effects are felt in the intermediate age groups, and, to the extent that birth rates are responsive to economic conditions, longer term changes in the younger age cohorts may occur. Since the birth rate linkages are generally weak and are excluded in this model, the principal impacts on total population occur through reductions in labor demand. In the experimental run the reference region population is reduced by only 81 and 86 persons in 1975 and 1980 respectively, representing losses of less than 0.5% compared to base run estimates; water restrictions of the magnitude included in this experimental run have very minor effects on aggregate population levels.

**Interindustry Sector**

Both direct and indirect effects on industry outputs will be expected from the water supply restriction. In this experiment the outputs of all manufacturing industries (3–9) were subjected to the 10% constraint so if gross output capacities are reduced in these industries indirect reductions in the remaining industries will occur. The results of the experiment indicate that gross outputs by industry were reduced relative to the base run in 13 to 15 industries in 1975 and in all 15 industries by 1980. Only the agriculture and mining and construction industries were unaffected by 1975. The three industries showing the largest negative impact on 1980 gross output included machinery, food and kindred, and transportation with declines of 39.7, 9.5 and 8.1% respectively.

The principal components of final demand were reduced slightly in both 1975 and 1980, reflecting declines in population, income and industrial investment. Personal consumption expenditure (PCE), purchases by the federal government (FGE), and capital accumulation (CA) are therefore indirectly influenced through restrictions on gross outputs, labor demand, income and new outlays for investment. These restrictions by industry tended to increase over the simulation period. Comparisons of water output capacity (OPW) and realized output (OP) in 1975 show that water supply was directly limiting on realized gross output for six of the seven industries where the constraint was directly operative and in six of eight industries in which it was not restricted. By 1980, agriculture, mining and construction, and machinery had also encountered this limitation. The fact that the machinery industry showed the largest reduction in gross output but was among the last sectors to encounter a direct capacity limitation attests to the importance of the indirect effects on the expansion of final demand and industry capital formation. Had these indirect effects not occurred, water supply restrictions could have been expected to limit capacity output in only the constrained industries.

**Capital Sector**

Though investment expanded in all industries through 1980, aggregate investment declined substantially relative to base run estimates, and important selective impacts are found among industries. Total investment (IVST) maintained base run levels through 1970, but declined by 5.98% (1975) and 5.92% (1980) compared to the unconstrained estimates. As a result, capital stock estimates (KAP) are reduced relative to base run levels in 13 of 15 industries by 1975 and all industries by 1980. Investment by industry is uniformly lower in the experimental run compared to the base run in 13 of the 15 industries in both 1975 and 1980. The largest percentage reduction relative to the base run is found in the machinery industry at 57.7% (1975) and 43.4% (1980).

## Water Sector

As in the interindustry sector, substantial indirect effects are found in the water model. Intake requirements for industries 3–9 rise at rates equal to or less than the direct constraint over the simulation period and by 1980 the implied intake requirements for all industrial sectors is reduced. Total intake requirements for the reference region are reduced by 3.4% (1975) and 5.6% (1980) compared to base run estimates. This result suggests a possibly more significant indirect impact than might actually occur because of the inability of an industry model to depict input substitution which would result from relative price and technological changes. Similarly these impacts may be overstated if the general interdependence of the reference region is substantially less than is shown in the interindustry sector or firms are able to substitute imports for locally produced intermediate products where their direct and indirect water content is high. Also the proportional reductions among industries will tend to overstate the impact of water restrictions compared to a rationing scheme that maintained equi-marginal water values among industries. However, the marginal product schedules associated with linear industry production functions in the interindustry model are not directly estimable from the structural specification used here.

## Labor Sector

Reductions in labor demand (RLF) and supply (ALF) relative to base run estimates are estimated in both 1975 and 1980. The supply reduction is only 50 workers and seems consistent with the minor impacts on the aggregate population noted above. Labor demand fell by 105,197 workers in 1980 relative to base run estimates; the machinery industry alone accounted for 66,066 of this reduction. Despite the substantial decline in the demand for labor, however, both the base and experimental runs indicate an excess of labor demand relative to supply. In 1980 the demand for full-time employees exceeded supply estimates by 11.9 and 4.4% in the base and experimental runs, respectively. An implied labor shortage of this magnitude could easily be accounted for by slightly higher than projected rates of labor force participation and/or worker productivity.

### Income Sector

The effects on per person income due to the water restriction depends on the relative rate of decline in gross labor income and employment. Relative to the base run all gross income measures were reduced over the experimental simulation period. A comparison with the base run indicated that personal income per capita was reduced by $660 (or 15.3% in 1975) and by $339 (or 7.0% in 1980). Income per employed person, however, was actually higher in 1980 under the water supply restriction both for the entire reference region and in seven of eight occupational groups. Thus, a restriction on water supply was followed by reductions in required labor that were disproportionately larger than the similarly induced reduction in gross income.

### SPATIAL CONTEXT

The bases for spatial disaggregation provide important information on the variability in economic indicators among alternative regionalizations and the types of delineation problems deemed to be of particular significance. Indeed the areal basis on which data is presented probably constitutes the most important determinant of how socioeconomic problems are perceived. Due to the enormous amount of spatial detail that is estimated by the model, only summary measures are presented in this section. The means and coefficients of variation for a set of impact and development indicators in both 1965 and 1980 provide evidence as to their spatial variability and their convergence or divergence over the simulation period.

Comparisons of economic indicators among alternative regional units are themselves subject to problems of spatial interpretation. The comparison of an indicator of services availability, for example, may be specious unless the areas are delineated in a manner permitting the residents of each region approximately equal access to them. Among regions the costs of acquiring services may vary substantially though the measured per capita availability index may suggest otherwise. Thus the contrast of hydrologic river basins with

economic areas may be misleading if the disparate size of these regions directly influence the development indicator of interest. Water
supply and effluent impacts are, for example, more realistically
viewed at a specific point within the hydrologic unit.

The means and coefficients of variation are calculated for a variety
of indicators in both 1965 and 1980; these are more fully tabulated
in Fullerton [16, Tables 29 and 30]. The terms impact and development indicators differentiate gross magnitudes from relative proportions or per captia values. Table XXI summarizes the high-low
coefficients of variation (and their mean values) for 1965 and 1980.
Gross (undeflated) dollar values are in tens of thousands of dollars
and population or employment variables are in actual numbers (impact indicators) or proportions (development indicators). A consistent characteristic for both sets of indicators is a general tendency
to diverge over the period 1965–80. For example, the 1965 range of
the coefficients of variation for impact indices is from a low of 31.5
(personal income) among EA to a high of 158.7 (gross output)
among CDA. The range increased for 1980 and a similar pattern
was found for the development indices.

**Impact Indices**

RBH and RBE show a close correspondence in the average values
of the indicators, but RBE is the only regional unit not having personal income as the lowest coefficient of variation in 1965. Gross output and females aged 20–44 show the highest coefficients of variation
in each year; the RBE coefficients consistently exceed those for
RBH even though their mean values are equal. Among RBE males
aged (65+) showed the lowest relative variation in 1965 and the
average coefficients for RBE tended to exceed those for RBH.

Since increases in the average coefficients are expected for smaller
regional units (if the variances remain constant), EA and CDA
should have larger high and low coefficients relative to RBH and
RBE. As shown in Table XXI EA had the lowest coefficient of variation (31.5 for personal income) in both 1965 and 1980 for all alternative regionalizations. Similarly the low coefficient for CDA in 1965
ranked second among the lowest coefficients of variation in that year.
This evidence suggests that labor market delineations provide a relative self-contained, homogeneous unit within the reference region.
Since EA are functionally independent it also seems likely that distributions of income are probably quite similar among these regional
units. The addition of a resource criterion in RBH and RBE tends to

Table XXI

Impact and Development Indicators Ranked by High-Low Coefficient of Variation (Mean; Coefficient of Variation) for RBH, RBE, EA and CDA, 1965 and 1980

| Indicators | | RBH | RBE | EA | CDA |
|---|---|---|---|---|---|
| **Impact** | | | | | |
| | highest | gross output ($241,911; 68.3) | gross output ($241,911; 81.7) | gross output ($84,342; 95.1) | gross output ($25,688; 158.7) |
| 1965 | | | | | |
| | lowest | personal income ($127,159; 40.7) | males (65+) ($28,596; 69.8) | personal income ($47,684; 31.5) | personal income ($18,863; 39.4) |
| | highest | females (20-44) ($81,885; 83.0) | females (20-44) ($81,885; 95.3) | females (20-44) ($30,706; 131.5) | females (20-44) ($13,997; 212.5) |
| 1980 | | | | | |
| | lowest | personal income ($244,421; 40.7) | personal income ($244,421; 66.3) | personal income ($91,658; 31.5) | personal income ($36,663; 80.2) |
| **Development** | | | | | |
| | highest | per capita income ($3,150.3; 40.0) | per capita income ($3,571.0; 53.3) | per capita income ($3,341.6; 44.0) | primary employment (7.3; 34.3) |
| 1965 | | | | | |
| | lowest | males (0-19) (18.8; 5.5) | service employment (54.5; 5.8) | males (0-19) (18.9; 5.7) | clerical workers (16.3; 7.1) |
| | highest | per capita income ($6,653.5; 67.4) | per capita income ($8,055.7; 83.9) | per capita income ($7,634.4; 63.2) | primary employment (8.1; 56.1) |
| 1980 | | | | | |
| | lowest | service employment 61.4; 7.2 | service employment (60.8; 5.8) | males (0-19) (15.8; 5.6) | service employment (69.3; 4.9) |

obscure this functional independence though personal income remains the lowest coefficient in the case of RBH. Also it is noted that the average coefficient for EA ranked between RBH and RBE in 1965 providing further evidence as to the among-region homogeneity of this unit. By 1980, however, EA's average coefficient of variation had risen to third among the four regionalizations, suggesting some decline in the functional independence of this unit over time at least as estimated by the model.

Several of the consistently variant characteristics warrant further comment. The high variability for females aged 20–44 among all regionalizations may be explained by the extent to which services and residentiary type industries tend to be concentrated in the larger populated centers within the reference region. Since these activities provide the major employment opportunity for young females a delineation scheme that resulted in an unequal distribution of population centers among areas would produce a substantial variation in this variable. Since services tend to be hierarchically correlated with the size of city, regional units including SMSAs may differ substantially from the smaller labor markets. In 1965, EA-8 (Des Moines) had 16.5% of its population in the female age group 20–44 while EA-14 (Creston) had only 12.7% of its population in this age-sex cohort. Similar influences probably affect the relative variation in gross outputs among regions since high output producing industries tend to be localized near prominent population centers.

The consistency with which the same variables appear as lowest or highest in their variational characteristics among regions attests to the spatial homogeneity of the reference region. The extremely variant characteristics tend to hold up under alternative aggregations of the smallest regional units despite the exceptions noted above. Though the regionalizations will have distinctly different planning orientations and uses, Table XXI suggests the degree to which the projected characteristics by regional unit are mutually consistent.

### Development Indices

The development indices shown at the bottom of Table XXI depict a similar set of differences among and within regions. The coefficients of variation ranged from a low of 4.9 (percentage of total employment accounted for by services, CDA, 1980) to a high of 83.9 (per capital income, RBE, 1980). As with the impact indicators the means and coefficients of variation tend to rise over the period 1965–80 where the same variable constitutes the high or low coefficient in each year. This reflects the increasing spatial diversity of the reference region economy that is attributable to the economic growth estimated over the full simulation period.

Proportional or per capita differences tend to vary less than gross magnitudes among regions in the reference economy and the high/ low coefficients for particular variables show a less consistent pattern. Demographic and employment characteristics show the least relative variation among regional units, and per capita income is the most variant indicator except among CDA regions. The high variation in per capita incomes coupled with the low variability in gross personal income noted above suggests an inverse correlation between gross income and population. This would appear to be consistent with relatively high birth rates and low incomes in rural EA and the opposite relationship in the more urbanized regions. Though per capita income varies considerably among units the coefficients for EA are low relative to other regions, confirming the expectations discussed by Fox and Kumar [15].

Since mean or proportional values vary less among regions of alternative sizes, the coefficients of variation should be relatively independent of the region's scale. RBE showed the highest average coefficients of variation and the least variation is indicated for CDA. Though the type of development investment in these two regions will differ, water resource planning should be particularly sensitive to the greater variations in per capita incomes at the river basin level. The model's projections suggest that this will be an increasingly important consideration given present (base run) trends in the reference region. The relatively consistent ranking of the coefficients of variation by regional size even under proportional or per capita criteria suggests the greater importance of interbasin decisions regarding water resource development planning than for the smaller regional units distinguished in this study.

# 6

# SUMMARY AND CONCLUSIONS

## PROCEDURAL SUMMARY

In this study a six-component recursive simulation model is developed and empirically estimated. The model is designed to provide comprehensive social accounts for several types of regional units simulated through time, in keeping with the objectives of accounting consistency among all spatial units and detailed spatial disaggregation of information at the reference regional level. The principal procedures used in constructing the model may be summarized as follows:

(1) The construction of regional models should be based on the best information regarding the data needs of numerous planning agencies. Identification of the principal submodels and their detailed sectoral components relied heavily on consultations with prospective users. These state agencies included the Natural Resources Council, Employment Security Commission, Department of Health, the Coordinating Group for Water and Related Land Resources Planning, and the Interagency study groups of the Upper Mississippi

and Missouri River Basins. The diversity of interests present among these agencies attests to the usefulness of comprehensive social accounts that provide for common data needed in their respective planning functions. The spatial base for the information extends the model's usefulness to municipalities, school districts, county governments, multicounty transportation planning agencies and numerous resource-oriented governmental bodies. Many of these units are exactly or closely contiguous with the spatial regions distinguished in the present version of the model. The four derivative submodels (capital, labor, income and water) usefully supplement the basic statistical accounts that provide the age/sex disaggregated population data and the sectoral employment/output estimates.

(2) The spatial basis for selecting subareas relied on existing policy units and the aggregative consistency of the smallest regional building block. Counties are admittedly of declining political importance, but they exhaust the area of the state and may be readily disaggregated if they contain a major metropolitan area. Four areas were selected to illustrate the model's capabilities. Economic areas and critical demand areas are important spatial units both as economic and population concentrations within water basins and as planning jurisdictions for municipal and transportation services. Hydrologic and economically-defined river basins are directly useful in coordinating water resource investments along major streams and their tributaries.

(3) Regional simulation models must incorporate existing empirical information within a diverse body of theoretical constructs if they are to provide useful planning data. Though theory should provide guidelines for the collection of data, inconsistencies are often present, and the separate specification of six submodels in part reflects this problem. The estimation of intersectoral linkages was confined primarily to the determination of net migration rates for all labor active population cohorts, and shift-share analysis applied to employment in standard areas over the period 1950–1960; the latter provided information for estimating changes in regional share coefficients that form the basis for spatial disaggregation. The state input-output model was estimated from secondary data sources, namely the 1958 interindustry model as developed by the Office of Business Economics and extended by the United States Department of Agriculture. Earlier studies of Iowa and the national economy were used to supplement informational gaps needed to complete the estimation of all intersectoral flows.

(4) Calibrating the model's output with observed data is a

frequently neglected, but important, procedure in estimating regional models. Because of some data limitations in a model of this size, verification was confined primarily to the demographic and labor sectors. Population census data for 1970 and employment estimates based on establishment surveys for 1967 provided the basis for checking the model's predictions. The principal problems included an inability to estimate net migration for college age cohorts in standard areas containing major educational institutions and an overstatement of expected disposable income per capita and capital accumulation when gross investment was based solely on a history of industry gross outputs. Modifications were introduced to calculate net migration for the two standard areas, and bounds were placed on industry investment rates as in Maki, et al. [37]. A final procedural step was accomplished by a rerun of the model for the period 1961–70 and comparisons to further data available for the region. A base run incorporated all modifications and was extended to 1980; an experimental run incorporating a 10% decline in implied water intake requirements for all manufacturing industries was utilized to compare with base run estimates.

(5) From the basic data estimated for subareas an array of development and impact indices were estimated. These indices monitor the development status of the various subregions at different points in time and provide evidence regarding the among and within region homogeneity of the various characteristics. Comparisons of these attributes are summarized by calculating coefficients of variation from the data estimated for 1965 and 1980.

## SUMMARY OF RESULTS

The verification procedures discussed above confirmed the model's capability to estimate closely population and employment over the period 1967–70. Comparisons of estimated total population in 1970 indicated divergences between model output and census data of less than 1.0%. Differences between model output and census figures varied by as much as 25.8% in some small regions, but with the exception of two cases these variations were less than 12.0%. In 1967 the employment comparison suggested that the model's estimate for the reference region was lower than an independent study estimate by 4.4%. Individual industry projections differed by less than 10.0% in all but one industry, and comparisons to 1970 estimates showed results of a similar magnitude.

The base run comprised a rerun of the historical phase (1961–

1970) and its continuation for the years 1971–1980. If the model's assumptions are met, reference area population can be expected to increase to 3,020,637 by 1980, or an annual average increase of 0.5%. Between 1965 and 1980 the female population declined in 7 of the 16 age groups, while the male population declined in 9 or 16 age groups. Uniform declines were noted for both males and females in the first four cohorts, while increases were common in all cohorts greater than 64 years of age.

Changes in the output and capital stock variables were markedly positive over the period 1965–1980 and may be attributed to increases in reference region population and income. Comparisons of OP and OPD showed that the latter was limiting in 11 of 15 industries in 1965 and all 15 of the industries in 1980. Trends in realized output show increases for all of the industries between 1965 and 1980. Similarly, changes in the capital stock were uniformly positive for all industries over this period; the total capital stock was estimated to increase by 59%.

Employment and gross output have proportional impacts on changes in water intake, use, consumption and discharge. Substantial increases in intake requirements occurred in the construction, machinery and finance, insurance and real estate industries; respective increases for these sectors are 88, 148 and 93% over the period 1965 to 1980. Employment demand was estimated to increase by about 295,000 workers between 1965 and 1980, which represents an annual average increase of 1.6%. Both aging and a shift to a heavier reliance on female workers was projected. The percentage of workers found in age groups 18–24 and 25–34 increased from 13.5 and 14.3% in 1965 to 17.2 and 18.9% in 1980. Female workers accounted for 27.8% of the labor force in 1965 and 38.4% in 1980. These increases in employment were accompanied by a 77% increase in per capita income over the projection period.

The experimental run projected minor differences in population due to water supply restrictions but substantial impacts on the interindustry sectors. Realized outputs were reduced by the water limitation in 13 of 15 industries by 1975 and in all 15 industries by 1980. The largest negative percentage changes were in the machinery (39.7), food and kindred (9.5) and transportation (8.1) sectors. A feedback impact was noted on the PCE, FGE and CA components of final demand in both 1975 and 1980.

The results of calculating measures of spatial dispersion for the regional impact indicators generally showed that the variation in these indices is inversely related to area size as measured by population. The measures of relative variability for population magnitudes were uniformly lower in EA than in RBE, but higher than in RBH; CDA showed the greatest relative variability for all impact indicators. Examination of the coefficients of variation for the development indicators revealed considerable differences among indices that could be attributed to the arbitrary choice of spatial unit. Coefficients of variations for per capita income ranged from a low of 40.0% among RBH to a high of 53.3% for RBE and for primary employment from 24.5% in RBH to 34.6% in EA's. Further it was noted that among RBH, RBE and EA, coefficients of variation were largest in EA's for 13 of the 21 indicators considered. This result provides some evidence of the sensitivity of labor markets to commonly used development measures and suggests the heterogeneous effects among these spatial units that may be expected from alternative patterns of growth.

## EXTENSIONS AND CONCLUDING REMARKS

The use of simulation models provides the best analytical format for extending the spatial and informational base of social accounts. While an impressive degree of generality may be achieved, time and cost limitations often preclude more extensive experimentation with the model's structure and alternative sources of informational input into the computer simulations. The experience attained in this study suggests the following principal extensions of the present work.

### Structural Limitations

Deterministic intersectoral linkages preclude a more realistic treatment of random shocks within most regional models. This problem is more serious where important structural equations have been either excluded or misspecified and the reference economy is subject to substantial exogenous influences of a nonregular nature. As models become more general, random shocks in specific equations should be less of a problem since more causal linkages are presumably estimable, but extensions could usefully include randomly drawn residuals for the principal regression linkages in the model. These modifications would add an element of realism for example in regression explanations of birth rates where substantial yearly changes have occurred. In equations where heteroscedasticity is a problem such adjustments may be particularly important.

The reliance on linear specification in the principal matrices and equational relationships may be unrealistic and fails to exploit the capabilities of simulation techniques in using equational systems whose analytical solutions are either unknown or difficult to derive. Where nature imposes fixed limitations as in the determination of age-specific fertility rates, asymptotic equations may be specified and tested. Properties of stability in the principal development indicators may be of particular interest where extensive non-linear estimation is used. Though the problem of spatial aggregation may be complicated where the objective of the analysis is essentially a system of linear and additive regional accounts, additional realism may be introduced by experimentation with non-linear equational systems.

Optimizing models that utilize programming techniques can be a useful supplement to simulation either in its dynamic form or for specific years of the computer run. The dynamic programming model can be particularly useful in studying public investment programs, and the regional allocation of factor inputs could be monitored and compared to simulated allocative patterns. The shadow prices of specific factor inputs are of substantial interest to public and private decision-makers and provide useful information in guiding the model-builder in making alternative assumptions regarding the desired level and rate of change in public expenditure programs.

Qualitative and quantitative limitations in essential data inputs are another problem in constructing large scale regional simulation models. The data inputs are assumed to be free of sampling and measurement error; the substitution of proxy variables may lead to incorrect interpretations of the model's output. Few guidelines are provided in estimating sampling error from governmentally constructed series (particularly below the federal level) and measurement error may be quite high even in census publications. The compounding of these errors due to the highly interrelated nature of simulation models is a subject that has yet to receive even cursory recognition and analysis. Also the estimation of regional magnitudes by comparisons to reference area data presupposes a regional structure of consumption preferences and productivity that are implicitly fixed in regional models. The implications of using alternative estimators as a basis for location quotients (employment, earnings, gross sales, etc.) have not been intensively examined. Given these data problems the model developed in this study suggests that more research is needed on the optimality of alternative schemes of spatial aggregation and disaggregation. Two alternatives are either building up from the smallest units or disaggregating down from the largest region. The former method suffers from an underdeveloped data base but only requires additivity. The data base is usually much more developed for larger units but errors of disaggregation may occur as some method of consistently breaking down the regional control totals is necessary.

## Sectoral Limitations

Within the demographic sector, the estimation of fertility, death and net-migration rates are particularly important problems. In our model these rates for age cohorts 13–16 were based on 1968 data and did not change throughout subsequent years. The recent changes in birth rates in Iowa suggests that additional work is needed in explaining age-specific fertility, and alternative migration models such as gravity and Markov formulations have not been tested within simulation models of this type.

Several data recommendations are a result of estimating problems in the demographic sector. The first concerns the availability of information on female fertility in substate areas. It is currently necessary to assume that the proportion of total live births attributed to fertile age groups in each substate area is exactly proportional to the same figure at the state level. Thus given data on numbers of fertile females and live births by area it was possible to calculate a unique fertility rate for each substate region. Only a minor change in the information categories and format used in compiling vital statistics at the state level could rectify this problem. The second problem is due to the necessity of using the residual method of calculating net migration by substate area. It is apparent to many government officials and regional economists that much improved data on gross migration could be obtained by school district in most states with extensive school aid programs. Though school districts do not coincide with counties, these data might be useful in estimating the gross components of migration because the latter are usually of greater interest than the net estimates. The addition of more accurate population data would be useful in estimating a full demographic accounts model for substate regions as is provided by Stone [48].

Several problems and extensions should be noted in the interindustry and capital sectors. The major difficulties in estimating subnational flows tables are well known, but it is likely that most state input-output models insufficiently utilize primary data available at the state level. In states with retail and corporate income taxes it is likely that these data could provide improved estimates of the size of domestic sectors and imported inputs. Two possibilities for extending this sector have also been suggested by Barnard [6] and MacMillan [28]. Their models expanded final demands and primary inputs of the processing· industries to generate social accounting matrices, and MacMillan suggested a means for extending these matrices to an exhaustive set of substate regions. Limitations in the capital sector arise from an insufficiency of data on capital stock and investment for subnational regions; the prospects for obtaining better information in this sector are not good.

Data limitations are also an important problem affecting both the demand and supply estimates in the labor sector. On the supply side, estimates of the available labor force are dependent on national trends in labor force participation. Labor demand projections for substate areas rely on the accuracy of regional share coefficients used in disaggregating required labor from the reference region to standard areas. Analysts have encountered numerous statistical difficulties in estimating share coefficients, and in this study variables assumed to be independent yielded less than satisfactory statistical results for several industries. The estimation of regional production functions as suggested in Borts and Stein [10] would be a useful extension of work on the demand for labor. The use of national trends in labor productivity also requires strong assumptions regarding rates of technological diffusion, the mobility of inputs among regions and the determinants of investment. Extensions of the labor sector should also include a means for assessing qualitative characteristics of the labor force.

Alternative estimation procedures could be usefully contrasted in the income sector. Commonly used bases such as industry or class of worker wage rates were discarded in favor of average earnings by occupational group; the latter are available for the reference region and are assumed to hold at the area level. Comparisons of relative earnings over time revealed unusual stability, and it seemed reasonable to assume that the variability in earnings among subregions was no more than would be expected in other measures of the return to labor. However these assumed regularities should be tested and an extension of this sector could include social accounting matrices for the primary input rows of the input-output model.

Finally, qualitative problems are of increasing importance in the water sector. Salinity, alkalinity, turbidity, oxygen demand and nitrogen content are but a few dimensions of water quality that are currently difficult to incorporate into models of this type. Point and time-specific data on water discharge normally does not include information on its qualitative characteristics. Further research will be necessary to estimate accurately the water loss incurred between the productive unit and the water system that is being analyzed; even the gross pollutant loads for specific subareas are not estimable in a comprehensive manner. The qualitative requirements of water as an input should be studied in more detail as a part of a general water quality accounting system for each industrial sector. Such an effort would necessarily involve the estimation of linkages to the demographic, interindustry and labor sectors.

## Policy Issues and Concluding Remarks

As suggested in the introductory chapter there are numerous policy issues in regional development that models of this type should be capable of incorporating. Many of these issues require vastly more detailed data bases than are currently available and structural adaptations within the sectors discussed above. Several of these issues and their problems should be noted.

Land use planning must be based on substantially improved accounting systems that can estimate the land-absorbing requirements of residentiary and producing sectors. The production and employment calculations in our model are readily converted to land use estimates except for the absence of good data on private land use. Since land requirements vary by size of firm and household, models must be adapted to handle the problem of size distributions of these units. A beginning could be made by differentiating size constraints by predominantly urban and rural areas within the model. Rural areas typically have the smallest size of productive unit in all economic sectors and the largest average family size; the opposite is apparent in metropolitan communities. Capacity expansions at existing plants and entirely new firms must similarly be distinguished if land use patterns are to be accurately estimated. This may require substantially more detailed data on the specific products that are fabricated (or services offered) and existing rates of plant utilization.

Parcel contiguity and the specific relation of one land use to another must also be incorporated within the model. Externalities in land uses are a crucial justification for planning and can never be estimated from regional control totals. This is not inherently a problem of spatial disaggregation alone for even a model which distinguishes lots in urban communities would not necessarily be capable of dealing with this problem. The locational context of all parcels of land to each other must be treated and their socioeconomic interdependencies should be specified. Even the largest metropolitan planning models have not yet incorporated these spatial interrelationships. The development of a land use sector would also allow for the more detailed estimation of the origins and destinations of residential and commercial trips as a basis for transportation planning.

Various environmental issues could also be incorporated into the present model with the aid of an expanded data base. Waste disposal requirements are directly related to levels of economic and residentiary activity and could be projected much in the same way as water requirements are estimated. Gross air pollutant loads could also be calculated from data relating sectoral outputs to particulate, sulfur dioxide and sulfur trioxide levels. Estimates for both of these problems would be improved if a finer degree of sectoral disaggregation could be obtained in producing sectors.

The recent energy crisis suggests other problem areas for simulation models of regional economies. Gross energy requirements are also related to the model's projections of economic and residentiary activity so calculations comparable to the waste disposal and air pollution estimates could be made; a finer degree of sectoral disaggregation would again be desirable. The locational distribution of activities is also important for estimating the energy requirements of the commercial transportation sector and the use of energy in bringing households to and from employment sites. Trip distribution models could be incorporated to account for these phenomena.

Problems of human resource planning necessitate an even finer array of characteristics estimated within the demographic sector. The stock of skills and educational attainment would be a useful supplement to data on occupational requirements. Mobility among and within regional units complicates the problem of providing gross estimates since characteristics of inmigrants are usually unknown. Rates of family formation, divorce, mental incapacity and physical disability are examples of sociological variables that have important implications for programs in the human resource field. Models that can estimate these many sociological characteristics of families and individuals over the life cycle and perhaps even carry individual household units through time have yet to be developed.

Natural resource conservation is a final area where regional models could be usefully applied or extended. Characteristics of animal and plant populations within regions and their relation to residentiary and productive activities have not been extensively studied. Problems of recreational demand and the depletion of forest preserves within the context of large scale computerized models are just beginning to be investigated. The feedback effects of policies that seek natural resource preservation and conservation are little appreciated and understood within the context of regional development.

These issues provide numerous challenges to the regional modeler at a time when the potentials of computer-estimated projections models are just beginning to be realized. Admittedly, problems with these methods are numerous. Currently fashionable policy issues rarely are estimable without a period of intensive data development. The interacting linkages are numerous and the best theoretical concepts may not be adaptable to available data. The user may not understand the structural basis for the estimates or the planning implications of the array of projected indicators. Administrative methods of effectively utilizing models on a continuing basis have not been successfully developed, and it may be some time before the planning profession adapts its curriculum to the demands of a more technical education necessary to interpret these data.

Despite these limitations large scale simulation models provide a flexible and comprehensive tool for analyzing numerous sectoral interactions in regional economies. Though the objectives of comprehensive planning may never be fully realized *de facto,* a system of spatially and temporally consistent and rapidly computable accounts that provide an empirical base for the decisions of planning agencies can do much to achieve this goal. Despite currently pessimistic views, several characteristics suggest the value of these models in regional economic planning.

The diversity of economic planning functions among agencies obscures the commonality of information required for decision-making. School districts, multicounty planning agencies, the U.S. Corps of Engineers and the many other planning units discussed above all require projected population estimates. Though the type and degree of cohort disaggregation may vary, the demographic needs of most important planning agencies are closely approximated by the type of model discussed in this study. Supplementary sectors provide causal linkages to study the stability of projected estimates in addition to a wealth of information on the subregions of the state. An office of statistical projections utilizing a single simulation model could achieve substantial savings for the numerous agencies currently conducting individual studies.

Consistent spatial estimates reduce the costs of reconciling conflicts among planning agencies. Where alternative data inputs are possible, their use in a single model provides exact comparisons of the results of using different statistical sources. Comprehensive models are usually capable of incorporating numerous alternative assumptions, further reducing the potential conflict among agencies. The consistency of both subarea aggregates and reference region totals avoids the problems of credibility often found in individual agency-contracted studies.

Consistent temporal projections provide a common base for long run planning, capital budgeting, and current budget askings of planning agencies. Both horizontal and vertical compatibility among governmental units is attained, and interagency comparisons of requested programs are based on a common informational base. The costs of both preparing budgetary estimates and reviewing subagency requests could be reduced.

Effectively implemented the advantages of computer simulated social and demographic accounts are numerous. Certainly they provide the only means by which reasonable empirical approximations to the numerous socioeconomic complexities can be constructed. As a common basis for most important planning functions of public agencies, their further development and adoption provides the best prospects for attaining the objectives of comprehensive planning.

# Appendix

# ESTIMATION METHODS

Two specialized calculations are described briefly in this appendix. (1) The migration of dependent children requires assumptions regarding the survival rate of migrating children and the fertility experience of females in the net migration age groups. This discussion explains in detail the assumptions underlying Equation 8 in Chapter 3. (2) Income estimates may be provided in several ways as suggested in Chapter 4. The specific calculations underlying the use of occupational earnings data are described in the second section below.

## MIGRATION OF DEPENDENT COHORTS

Net migration of dependent children was determined in the demographic sector by evaluating the following equation:

$$(NM)^t_{rsa} = (SR)^o_{sa} * \sum_{y-a+3}^{9+a} 2.5 * (NM)^t_{ray} * [(FR)^o_{rsy-a} + (FR)^o_{rsy-a+1}]$$

The number of net dependent migrants is assumed to be directly related to the number of net female migrants. Thus, the number of net dependent migrants includes all surviving dependent children born to migrating females in years previous to the current year;

the fertility of migrating females and the survival rate of their children are the basic determinants of dependent net migration .

An example of the calculations necessary to estimate $(NM)_{rsi}$, the net migration of children in age group 0–4, serves to demonstrate this procedure. Children of ages 0–4 can be associated with fertile females in age groups four through ten since during the past five years children could have been born to the females in any of these groups. The potential population of dependents in age group 0–4 is obtained by taking the product of the fertility rate and the appropriate female population (net migrants) in all fertile age groups and summing over age groups four through ten.

Fertility rates used in this study are based on annual data, which accounts for the necessity of multiplying by five. The multiplicative factor (2.5) was used for the following reason. During an average five-year period a female has been in both her current and prior age group for 2.5 years. This also accounts for the inclusion of two fertility rates in this equation. The potential population was then converted to the surviving population by multiplying it by the appropriate five-year survival rate. The same calculation is used for dependent groups two and three with a modification in the range of female age groups considered. Dependent groups two and three are associated with female age groups five through eleven and six through twelve, respectively. The fertility rates were based on 1960 data, and in 1968 the fertility rates of that year were incorporated.

Given five-year fertility rates, a new rate should be associated with each age group of dependents at five-year intervals in the simulation run. For example, in 1960 the population of age group one should be based on fertility rates that prevailed between 1955 and 1959, the population of age group two on fertility experience between 1950 and 1954 and the population of age group three on rates experienced between 1945 and 1949. In 1965 the associations between populations and fertility rates should be as follows: (0–4), 1960–64; (5–9), 1955–59: (10–14), 1950–54. The same relationship for 1970 should be: (0–4), 1965–69: (5–9), 1960–64; (10–14), 1955–59. As data were not available beyond 1968, fertility rates experienced in that year were assumed to remain constant throughout the remaining twelve years of the simulation period.

## ESTIMATION OF INCOME

Total earnings for the eight major occupational groups were calculated from census data for the years 1950 and 1960. Estimates of total earnings were developed by multiplying midvalues in the class intervals of earnings per employed worker in each occupation class by the number of employed persons earning at that rate and summing their product. Proportions of this total earnings figure were estimated for each occupational group. These in turn were used to allocate estimates of state personal income as obtained from the Survey of Current Business [59] to occupations.

Table A.1 shows the calculations necessary to provide the income estimates shown in Chapter IV. Division by the number of employed persons in each occupation provided an estimate of average personal income per employed worker by occupation group. These estimates are shown in columns 1 and 3 of Table A.1. Income relatives were calculated by dividing elements in rows 1 through 8 in columns 1 and 2 by the average income o fthe professional-technical group. The result of this calculation is shown in columns 2 and 4 of Table A.1.

Table A.1

Average and Relative Income of Employed Persons
by Major Occupation Groups, Iowa 1950 and 1960[a]

| Occupation Group | : 1950 | | : 1960 | |
|---|---|---|---|---|
| | :*Average* :*Income* | *Relative Income* | :*Average* :*Income* | *Relative Income* |
| Professional and technical | $3,247 | 1.00000 | $7,249 | 1.00000 |
| Farmers and managers | 3,501 | 1.07826 | 6,439 | 0.88826 |
| Clerical workers | 2,176 | 0.67013 | 4,506 | 0.62160 |
| Sales workers | 2,609 | 0.80351 | 5,671 | 0.78231 |
| Craftsmen and foremen | 2,994 | 0.92207 | 7,074 | 0.97586 |
| Operatives | 2,498 | 0.76931 | 5,882 | 0.81142 |
| Service workers | 1,394 | 0.42932 | 2,560 | 0.35315 |
| Laborers | 1,690 | 0.52047 | 3,643 | 0.50255 |

[a]Based on Survey of Current Business income data for states after conversion to 1960 dollars.

# REFERENCES

1. Almon, C. Jr. *The American Economy to 1975* (New York: Harper and Row, 1966).
2. Andrews, R. B. "Economic Planning for Small Areas: An Analytical System," *Land Econ.* **39**: 143–155 (1963).
3. Andrews, R. B. "Economic Planning for Small Areas: The Planning Process," *Land Econ.* **39**: 253–264 (1963).
4. Ashby, L. D. "Regional Change in a National Setting," U.S. Department of Commerce Staff Working Paper in Economics and Statistics No. 7 (1964).
5. Ashby, L. D. "The Shift and Share Analysis: A Reply," *South. Econ. J.* **34**: 423–425 (1968).
6. Barnard, J. R. *Design and Use of Accounting Systems in State Development Planning* (Iowa City, Iowa: Bureau of Business and Economic Research, The University of Iowa).
7. Barnard, J. R. "Iowa's River Basins Preliminary Water Use Projections for Iowa, Portions of Missouri and Upper Mississippi Basins," unpublished mimeographed paper (Iowa City, Iowa: Bureau of Business and Economic Research, The University of Iowa, 1967).
8. Barnard, J. R. and Maki, W. R. "Demand, Productivity, and Interindustry Relations in the Iowa Economy, 1954–1974," unpublished research bulletin manuscript (Ames, Iowa: Department of Economics and Sociology, Iowa State University, 1965).
9. Berry, J. L. "Strategies, Models, and Economic Theories of Development in Rural Regions," U.S. Department of Agriculture Report 127, (1967).
10. Borts, G. H. and J. L. Stein. *Economic Growth in a Free Market* (New York: John Wiley and Sons, 1959).
11. Chenery, H. B. and P. G. Clark. *Interindustry Economics* (New York: John Wiley and Sons, 1959).

125

12. Ehemann, G. C. *The Construction of Personal Income Estimates for Counties* (Iowa City, Iowa Bureau of Business and Economic Research, The University of Iowa, 1969).

13. Forrester, J. W. *Industrial Dynamics* (New York: John Wiley and Sons, 1961).

14. Fox, K. A. "Functional Economic Areas: A Strategic Concept for Promoting Civic Responsibility, Human Dignity and Maximum Employment in the United States," unpublished manuscript (Ames, Iowa: Department of Economics, Iowa State University, 1969).

15. Fox, K. A. and K. R. Kumar. "Delineating Functional Economic Areas," in *Research and Education for Regional and Area Development* (Ames, Iowa: Center for Agricultural and Economic Development, Iowa State University, 1966) pp. 13–55.

16. Fullerton, H. H. *An Economic Simulation Model for Development and Resource Planning,* unpublished Ph.D. thesis (Ames, Iowa: Iowa State University, 1971).

17. Goldman, M. R., M. L. Marimont, and B. N. Vaccara. "The Interindustry Structure of the United States: A Report on the 1958 Input-output Study," *Survey of Current Business* 44: 10–29 (1964).

18. Hamilton, H. R., S. E. Goldstone, F. J. Casario, D. C. Sweet, D. E. Boyce, and A. L. Pugh. *A Dynamic Model of the Economy of the Susquehanna River Basin* (Columbus, Ohio: Columbus Laboratories, Battelle Memorial Institute, 1966).

19. Higgins, B. *Economic Development* (New York: W. W. Norton and Company, Inc., 1968).

20. Isard, W. *Methods of Regional Analysis* (Cambridge, Mass.: The Massachusetts Institute of Technology Press, 1960).

21. Johansen, L. *Public Economics* (Chicago: Rand McNally and Company, 1965).

22. Lange, O. "The Output-Investment Ratio and Input-Output Analysis," *Econometrica* 28: 310–325 (1960).

23. Leontief, W. W. "Factor Proportions and the Structure of American Trade: Further Theoretical and Empirical Analysis," *Rev. Econ. Stat.* 38: 386–407 (1956).

24. Lewis, A. W. *Economic Development with Unlimited Supplies of Labour* (Manchester, England: The Manchester School of Economic and Social Studies, 1954).

25. MacKichen, K. A. *Estimated Use of Water in the United States, 1950* (U.S. Dept. Int.: Geological Survey Circular 115, 1951).

26. Mackichen, K. A. *Estimated Use of Water in the United States, 1955* (U.S. Dept. Int.: Geologic Survey Circular 398, 1957).

27. MacKichen, K. A. *Estimated Use of Water in the United States, 1960* (U.S. Dept. Int.: Geological Survey Circular 456, 1961).

28. MacMillan, J. A. *Public Service Systems in Rural-Urban Development,* unpublished Ph.D. dissertation (Ames, Iowa: Iowa State University, 1968).

29. Maki, W. R. "Information, Data, and Research for Economic and Social Policy," *Proc. Symp. Stimulants to Social and Economic Development in Slow Growing Regions,* Banff, Alberta, Sept. 6 9, 1966, Vol. 2. pp. 84–102 (Edmonton, Alberta: Department of Agricultural Economics, University of Alberta, 1966).

30. Maki, W. R. "Input-Output Model of the Iowa Economy," unpublished Data (Ames, Iowa: Department of Economics, Iowa State University, 1967).

31. Maki, W. R. "Iowa Economic Trends: Employment, Output, Income and Related Data," supplements I and II, unpublished mimeographed paper prepared for the Iowa Office of Planning and Programming under a contract with the Iowa Agricultural Experimental Station (Ames, Iowa: Department of Economics, Iowa State University, 1968).

32. Maki, W. R. *Preliminary Economic Projections for Iowa, Portions of Missouri and Upper Mississippi Basins* (mimeographed) (Ames, Iowa: Department of Economics, Iowa State University, 1967).

33. Maki, W. R. *Projections of Iowa's Economy and People in 1974* (Iowa Agricultural Experiment Station Special Report 41, 1965).

34. Maki, W. R. "Regional Economic Development and Water," unpublished paper presented at the North Central Water Resources Research Seminar, Chicago, Illinois, March, 1968 (Ames, Iowa: Iowa State University, 1968).

35. Maki, W. R. "Regional Research on Community Development," unpublished paper presented at the Farm Foundation meeting, Chicago, Ill., May, 1968 (Ames, Iowa: Department of Economics, Iowa State University, 1968).

36. Maki, W. R. and R. Suttor. "Analysis for Area Development Planning," in *Research and Education for Regional and Area Development* (Ames, Iowa: Iowa State University Press, 1966) pp. 193–214.

37. Maki, W. R., R. E. Suttor, and J. R. Barnard. *Simulation of Regional Product and Income with Emphasis on Iowa, 1954–1974* (Iowa Agricultural Experiment Station Research Bulletin 548, 1966).

38. Martin, W. E. and H. O. Carter. *A California Interindustry Analysis Emphasizing Agriculture.* (California Agricultural Experiment Station, Giannini Foundation of Agricultural Economics, Giannini Foundation Research Report, 250, 1962).

39. Mattila, J. M. and J. P. Concannon. "A Study of ARA Designated Counties," unpublished paper prepared for the Area Redevelopment Administration (Washington, D.C.: U.S. Department of Commerce, 1964).

40. Meyer, J. R. "Regional Economics: A Survey," *Amer. Econ. Rev.* 53: 19–54 (1963).

41. Miernyk, W. H. *The Elements of Input-Output Analysis* (New York: Random House, Inc., 1965).

42. Mincer, J. "Labor-force Participation and Unemployment: A Re-

view of Recent Evidence," in *Prosperity and Unemployment,* R. A. Gordon and M. S. Gordon, Eds. (New York: John Wiley and Sons, Inc., 1966) pp. 73–112.

43. Missouri Basin Inter-agency Committee. "The Missouri River Basin Comprehensive Study," unpublished preliminary draft (Omaha, Nebraska: U.S. Army Corps of Engineers, 1968).

44. Moore, F. T. and J. W. Petersen. "Regional Analysis: An Interindustry Model of Utah," *Rev. Econ. Stat.* 37: 376-377, 1955).

45. Mullendore, W. E. *An Economic Simulation Model for Urban Regional Development Planning,* unpublished Ph.D. dissertation (Ames, Iowa: Iowa State University, 1968).

46. Murray, R. C. *Estimated Use of Water in the United States, 1965* (U.S. Dept. Int.: Geological Survey Circular 556, 1968).

47. Ohlin, B. *Interregional and International Trade* (Cambridge, Mass.: Harvard University Press, 1933).

48. Stone, R. *Demographic Accounting and Model Building with Special Reference to Learning and Earning Activities,* unpublished paper prepared for the Directorate for Scientific Affairs (Paris, France: Organization for Economic Cooperation and Development, February, 1969).

49. Stone, R., J. Bates, and M. Bacharach. *A Programme for Growth, (Part) 3, Input-output Relationships, 1954–1966* (London, England: Chapman and Hall, Ltd., 1963).

50. Strotz, R. H. and H. O. A. Wold. "Recursive vs. Nonrecursive Systems: An Attempt at Synthesis," *Econometrica* 28: 417–427 (1960).

51. Tinbergen, J. *On the Theory of Economic Policy* (Amsterdam, The Netherlands: North Holland Publishing Company, 1952).

52. U.S. Army Engineer Division, North Central. "Upper Mississippi River Comprehensive Basin Study of Water and Land Resources," unpublished draft no, 2 (Chicago: author, 1968).

53. U.S. Bureau of the Census. *U.S. Census of Population: 1960. General Social and Economic Characteristics, Iowa* (1962).

54. U.S. Bureau of the Census. *U.S. Census of Population: 1960. Number of Inhabitants, Iowa* (1960).

55. U.S. Bureau of the Census. *Census of Manufacturers: 1963. Area Statistics, Number 16, Iowa* (1966).

56. U.S. Bureau of the Census. *U.S. Census of Manufacturers, 1958. Industrial Water Use.* U.S. Bureau of the Census Bul. MC58 (1961).

57. United States Department of Commerce. *Survey of Current Business. 1963 through 1968.*

58. U.S. Department of Labor. *Population and Labor-force Projections for the United States.* U.S. Bureau of Labor Statistics Bulletin 1242 (1965).

59. U.S. Office of Business Economics. *Survey of Current Business 45,* (11) (1965).

# INDEX

# INDEX